# auto mechanics

# auto mechanics

## an introduction and guide

by A. Edward Little

*illustrated with diagrams and photographs*

FRANKLIN WATTS, INC. New York / 1974

Library of Congress Cataloging in Publication Data

———

Auto mechanics; an introduction and guide.

SUMMARY: A basic explanation of how the various parts of an automobile function.

1. Automobiles – Maintenance and repair – Juvenile literature. [1. Automobiles – Maintenance and repair] I. Title.

TL152.K528 629.2 73-14689

ISBN 0-531-02670-1


Printed in the United States of America

6 5 4 3 2

# contents

*introduction* 1

*the source of an automobile's power* 3

*the basic principles of the automobile engine* 7

*engine structure* 14

*operation of the valves* 17

*engine characteristics* 22

*the fuel system* 24

*the electrical system* 33

*lubrication* 43

cooling systems 46

exhaust systems and emission controls 50

two-stroke and diesel engines 53

getting the power to the wheels 56

driving and stopping — brakes 74

steering 82

suspension systems 89

future developments in automotive engineering 93

glossary of automotive terms 101

some facts about the automobile and the industry 105

for further reading 109

index 111

# auto mechanics

# introduction

Few people in the world today are not affected in some way by automobiles. Actually, for most people the automobile is part of a way of life; no longer a luxury, but a necessity for coping with the problems of day-to-day living.

The automobile is but one — although one of the most obvious — artifacts of our time, which has been called the Machine Age or the Age of Technology. Our economy is based to a large extent on the assumption that at least most of the machines in use will continue to operate. No one person can ever hope to understand everything about all the machines that we depend on, but hopefully if something goes wrong with any given machine someone can be found with the expertise to put it right. And this is certainly the faith of millions of automobile drivers, most of whom have not the foggiest notion of what makes the thing go.

For the average driver, knowledge of automotive engineering does not go beyond knowing that the key is turned to start the car, the shift lever is moved to D or the gears shifted into low, the gas pedal is pressed to make it move, the brake pedal pressed to stop it, and the steering wheel turned to make the machine go where the driver wants it to go. And, of course, the knowledge that, at frequent intervals, a liquid called gasoline has to be pumped into a hole in the machine is rather universal.

In our society it has been traditional for women to know nothing about auto mechanics (although in fact not many men knew very much about the subject either). But like many notions that are changing in the late twentieth century, the idea of auto mechanics as a man's province alone must change, too. For the fact is that anyone — man, woman, boy, girl — who gets behind the wheel of an automobile — for the safety of himself, herself, passengers, or pedestrians — should understand something about that machine. The automobile is a complicated mechanism; that is true. But auto mechanics does not have to be, and is not, a mysterious, incomprehensible subject. It *can* be understood, and understanding can also be fun, as the next pages will show you.

# the source of an automobile's power

The internal-combustion piston engine, which powers over 99 percent of the world's automobiles, is an extremely complex machine. The mechanics' rule of thumb — the simpler a machine is, the more dependable it is likely to be — seems to be contradicted by the success of this engine. Almost one hundred years of engineering effort have gone into making today's complicated automobile engine the dependable machine it is today.

The history of self-propelled vehicles goes back almost two hundred years to 1789, when a Frenchman, Joseph Cugnot, mounted a steam engine on a wheeled platform. The piston of the engine was connected directly to the driving wheel. This so-called "steam wagon," which was intended to move heavy artillery, could achieve a top speed of about three miles an hour.

Through the nineteenth century there were several attempts to make other "horseless" vehicles. Most were powered by steam, but one, made by Isaac de Rivaz in France in 1807, was powered by a piston that was moved by an exploding mixture of hydrogen and air.

The development of the gasoline engine as we know it today began in the last quarter of the nineteenth century. An Austrian named Siegfried Marcus made several gasoline-powered vehicles in the mid-1870s. His machines worked, but they were not a commercial success. Karl Benz, working in Germany, started to make vehicles powered by gasoline engines in the 1880s, and he actually managed to sell a few of them. His first vehicle was a tricycle powered by a single-cylinder gasoline engine. Gottlieb Daimler, another German, was also making gasoline-powered vehicles at about the same time. Daimler and Benz never worked together, but in 1926 the companies using their names merged to form the Daimler-Benz Company, which began to turn out the first Mercedes-Benz automobiles.

Daimler, Benz, and others such as Nicolaus Otto were intrigued with the possibilities of internal-combustion engines. Both steam and gasoline engines are heat engines. That is, a fuel is burned, and the resultant heat is converted to mechanical energy — the energy of motion.

The steam engine is an *external*-combustion engine. The heat from burning fuel is used to boil water, and the steam produced is used to push a piston enclosed in a cylinder, or a turbine. In the *internal*-combustion engine (as in today's automobiles), the fuel is burned right in the cylinder. The hot, expanding gases that result from the combustion push the piston. Early internal-combustion enthusiasts believed that an internal-combustion engine was basically more efficient than an external-combustion engine, since boilers were not required. That meant that internal-combustion engines could be lighter than steam engines. There were, however, many problems. Fuel and air had to be directed into the cylinder in the proper mixture and vaporization. Some means had to be provided to get rid of the waste gases. And a mechanism had to be devised for opening the cylinder to allow the entrance of fuel and the exhaust of waste gases, then closing the cylinder for compressing and burning the fuel. All of these requirements indicated that a working internal-combustion engine would be an extremely complicated device.

The development of automobiles began in earnest in the first decade of the twentieth century. But automobiles of this period tended to be expensive and unreliable. Even so, there were enough interested people with enough money to buy the new machines. All kinds of enterprises, ranging from one-man operations in barns to huge factories, become part of the automobile business. Hence, the pressure of competition forced rapid improvements in design.

The problems inherent in those early internal-combustion engines were evident, and so, many other kinds of power plants were tried. Steam-powered automobiles were very successful in the first quarter of the century. Electric-powered automobiles were also popular. Internal-combustion engines that burned fuels such as naphtha, alcohol, hydrogen, and other flammables had little success. There were even a few autos driven by windup clockwork.

Gasoline engines of the period were very noisy, but that was the least of the owners' problems. They sometimes exploded, and they required a great deal of maintenance. However, despite its problems, the gasoline engine eventually won out over steam and electricity. While steam cars were fast and reasonably dependable, they required time to get up a head of steam, and operating them was somewhat complicated. The potential motoring public, wanting instant mobility, did not have the patience to drive steam automobiles. As for electric cars, they were quiet and dependable, but the batteries had to be frequently recharged, and the best of them were good for only twenty or thirty miles on a charge. Their silence, however, contributed to their wide use by wealthy society ladies for short trips around the city.

Another factor that contributed to the acceptance of the gasoline engine was a series of well-publicized races from 1890 to about 1910. Almost all of them were won by gasoline-engine-powered vehicles. One of the more famous of the contests was the Paris-Bordeaux race of 1895. It was won by a Panhard-Levassor, powered by a one-cylinder gasoline engine designed by Gottlieb Daimler. It outlasted a field that included six steam cars, fifteen gasoline cars, and one electric car.

As the demand for automobiles increased, automotive engineers turned their talents to perfecting and refining the gasoline engine. Thus, the once noisy, cantankerous contraption that seemed to violate all concepts of what a good machine should be was eventually transformed into a dependable machine that could run at high speeds and temperatures for thousands of miles without breaking down.

The success of the gasoline engine has not been without its price, however. The exhaust of millions of such engines has created a serious air pollution problem, especially in cities. But just as the pressure of competition stimulated the development of the efficient engines that power today's automobiles, so surely will the pressure of law and a demanding public force the development of nonpolluting engines.

# the basic principles of the automobile engine

An old muzzle-loading cannon operates on the same basic physical principles as does the automobile engine. A powder charge is put into the cannon, which is basically a hollow cylinder. The powder is tamped down, or compressed, by a rammer, and a cannonball is put in place over the powder. The powder is ignited and it burns rapidly. The expanding gases resulting from the burning force the ball out of the cannon.

In the automotive engine, the *cylinder* can be compared to the cannon and the *piston* takes the place of the cannonball. The piston is not shot out of the cylinder but is moved in a controlled manner by the expanding gases formed by the burning of a mixture of air and gasoline. The up-and-down movement of the piston is the motion that ultimately turns the wheels of the automobile.

The cylinder is the basic unit of the engine. It is closed at the top except for two valves, an *intake valve* to allow fuel into the cylinder and an *exhaust valve* to allow exhaust gas to escape from the cylinder. The cylinder itself is somewhat like an open can with the open end away and down from the top of the engine.

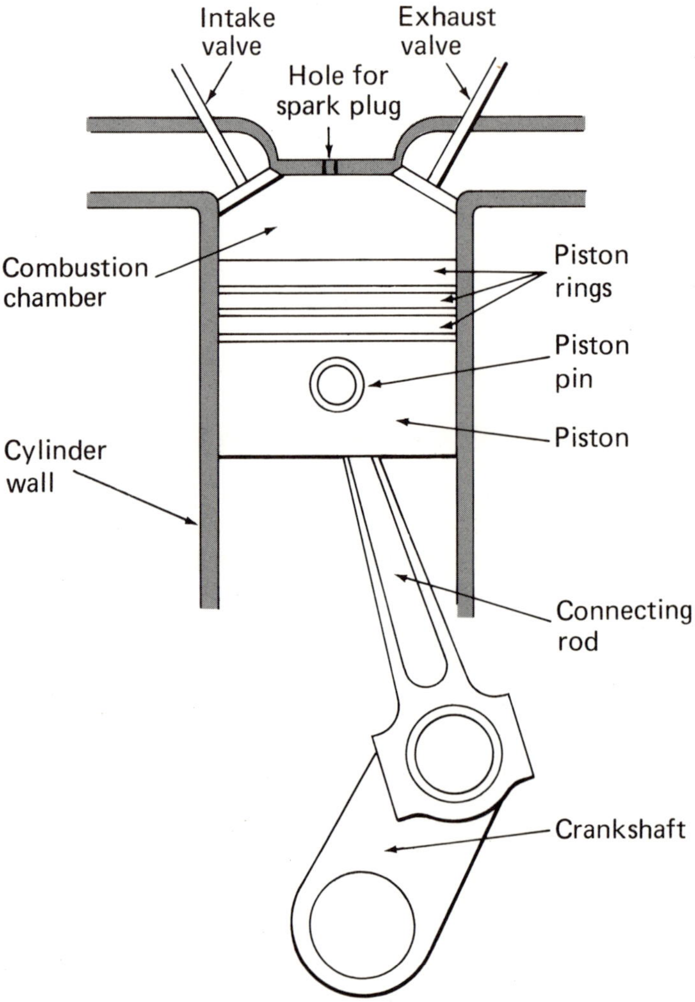


*The cylinder is the basic unit of the internal-combustion automobile engine.*

The piston is fitted inside the cylinder. The fit is tight, but not so tight that the piston cannot move. Most piston engines are four-stroke engines, so named because the piston moves four times for each power-producing explosion. The series of four strokes is called the cycle. (It is also called the *Otto cycle* after Nicolaus Otto, who devised it for use in stationary engines for powering various factory machines.) The four strokes are: (1) intake, (2) compression, (3) power, and (4) exhaust. This is how each works:

*Intake Stroke.* The piston moves down and the intake valve opens. The downward movement of the piston creates a vacuum, or suction, and the gasoline-air mixture is drawn into the cylinder.

*Compression Stroke.* The valves are closed and the piston moves up. The fuel-air mixture is compressed against the top of the cylinder. The more tightly the fuel is compressed, the more power is likely to be obtained in the next stroke.

*Power Stroke.* The valves remain closed. The fuel mixture is ignited by an electric spark. The rapidly expanding hot gases force the piston downward. The hot expanding gases have to go somewhere, and since the piston is movable, the power of the explosion is more rapidly expended against the piston.

*Exhaust Stroke.* The exhaust valve opens and the piston moves up, forcing the exhaust gases out of the exhaust valve.

Only one stroke in the four-stroke cycle produces power. An outside force has to be applied to the engine to get it started and a means must be provided to keep the engine going between power strokes. The electric spark and the opening and closing of the valves must be timed to coincide with the corresponding movements of the pistons. Correct timing is extremely important.

The piston must be slightly smaller in diameter than the cylinder, or *bore*. This is to allow for movement and expansion. The piston expands more than the cylinder when it gets hot. If the piston were too tight, the expansion might cause it to be so tightly jammed against the cylinder wall that it could not move. However, some kind of seal must be provided to keep the maxi-

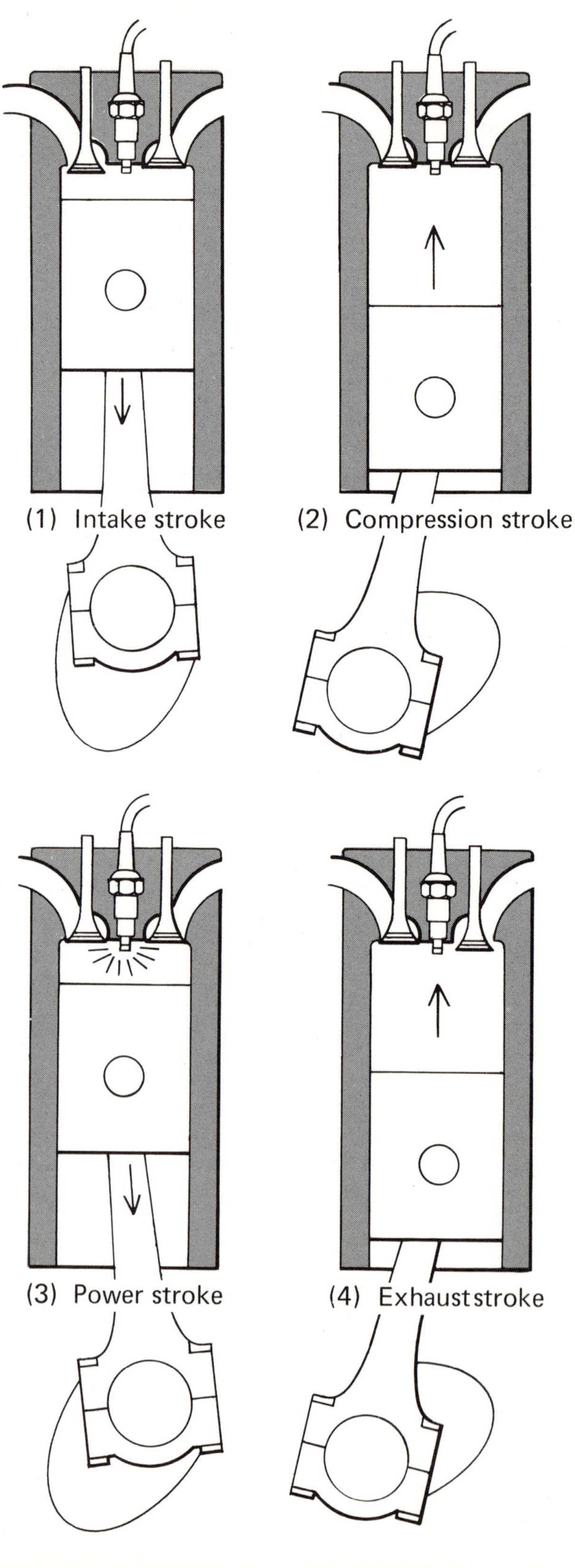
(1) Intake stroke
(2) Compression stroke
(3) Power stroke
(4) Exhaust stroke

mum force of the combustion inside the cylinder. The seal is provided by *piston rings*, which fit into grooves around the cylinder.

Piston rings have a diameter slightly larger than the cylinder, but a break in each ring allows it to be squeezed to fit in place. The rings are made of a particularly springy metal, which allows them to provide a snug fit while still allowing motion, and a special alloy, which lets them remain springy when hot.

The piston rings also control the amount of lubricating oil entering the interior of the combustion chamber. Oil is needed to lubricate the cylinder wall, but if too much oil gets into the combustion chamber, it will be burned and this will result in a smoky, greasy exhaust. Each piston is usually fitted with three or four rings. The bottom ring has vents that control the entrance of oil into the cylinder.

Most automobile engines have more than one cylinder. The usual number is four, six, or eight. In the past, automobiles with twelve or even sixteen cylinders were manufactured. In multicylinder engines, power strokes, or firings, occur in different cylinders at different times in a definite sequence. This tends to keep pistons that are not on a power stroke moving. A heavy *flywheel* at the end of the crankshaft (discussed below) also tends to keep the pistons going between power strokes. Once the flywheel is moving, its inertia keeps it moving. (Inertia is the tendency of an object in motion to remain in motion and the tendency of an object at rest to remain at rest.)

In the course of the cycle, the pistons move up and down. This up-and-down, or reciprocating, motion must be changed to rotary motion; that is, ultimately it must be transmitted to the wheels.

Each piston is connected by a *connecting rod* to a *crankshaft*. (See the illustration on page 15.) The crankshaft is basically a metal rod made in angled sections so that when the connecting rod pushes down on it the crankshaft is turned.

The connecting rod is made of forged steel and is somewhat tapered in shape, providing a circular small end and a big end. The small end is attached to the inside of the cylinder and the

*The four strokes of the piston engine*

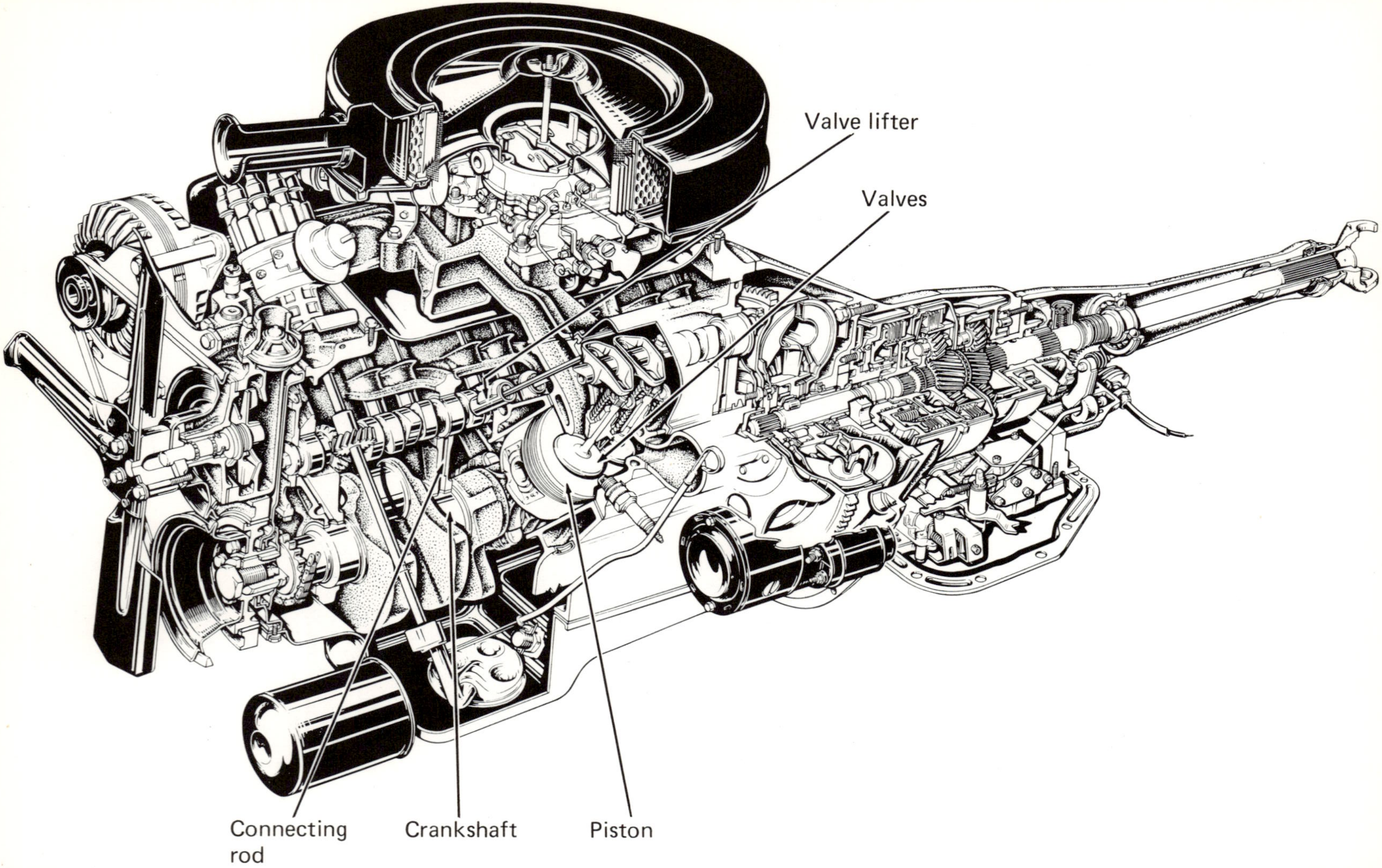
Valve lifter
Valves
Connecting rod
Crankshaft
Piston

big end is attached to the crankshaft. The small end is suspended on a *piston pin* that is inserted through holes in the center of the piston. The pin is supported by *bosses*, or clips.

This arrangement allows the connecting rod to swing as the crankshaft turns. The big end is secured to the crankshaft at the *crankpins*. Between the crankpins on the crankshaft are the *main bearings*. The crankshaft is supported in the engine block by the main bearings. The flywheel is usually attached to the rear end of the crankshaft by means of a flange.

The cylinders can be arranged in line, one behind the other, or in a V. Most four- and six-cylinder engines are in line, while most eight-cylinder engines are V-type. There are, however, a few cars that have V-4 and V-6 engines, and "horizontally opposed" engines. The V arrangement, as well as the horizontally opposed, allows the engine to be shorter. It also tends to run more smoothly with less vibration. The cylinders in a V are usually opposed at an angle of 90 degrees, although there are some with 60-degree-angle oppositions. Horizontally opposed pistons are 180 degrees opposite each other. In V-8 engines, the crankpins are longer, to accommodate two connecting rod big ends, one from each opposing piston. The single most popular engine in use today is the horizontally opposed four-cylinder model powering the Volkswagen.

The pistons are not all fired at the same time, but in a predetermined sequence. If they were all fired at the same time, the car would tend to jump up and down, and the vibrations would be so great that the engine would come apart. Firing orders are chosen to minimize vibration. A typical firing order in a six-cylinder engine might be 1-5-3-6-2-4. Vibration is also minimized by counterweights in the crankshaft, which offset the weight of the metal in the crankpins.

The distance that the piston moves in one direction is the *stroke*. Different engines are designed with different strokes, but in recent years manufacturers have tended to make short-stroke engines. Short-stroke engines can be lower and the valves can be placed in more mechanically advantageous positions. The short stroke tends to reduce friction power loss.

*A four-barrel, V-8 engine*

# engine structure

The main bulk of the engine is the cast-metal *cylinder block*, usually referred to simply as the block. It is supported by, and attached to, the *chassis* (frame) of the automobile. The cylinders are bored into the block so that all the cylinders are in an enclosed unit. A *cylinder head* is placed over the block like an oblong inverted bowl. It is fastened securely to the block with bolts. The cylinder head forms the upper part of the combustion chamber, and in most modern engines the head encloses the valves. The head has threaded holes to accommodate spark plugs. In V-8 and opposed engines, there are two cylinder heads.

Since the cylinder head is part of the combustion chamber, the head and the block must be tightly sealed together. This is done by placing a *gasket* between the head and block. A gasket is a flat piece of flexible material, frequently asbestos. It is cut to conform to the shape of the block and the bolts, screws, and other fittings that must pass through it.

The *oil pan* is fitted to the lower part of the block. It closes off the block and holds the lubricating oil. (It is also called the

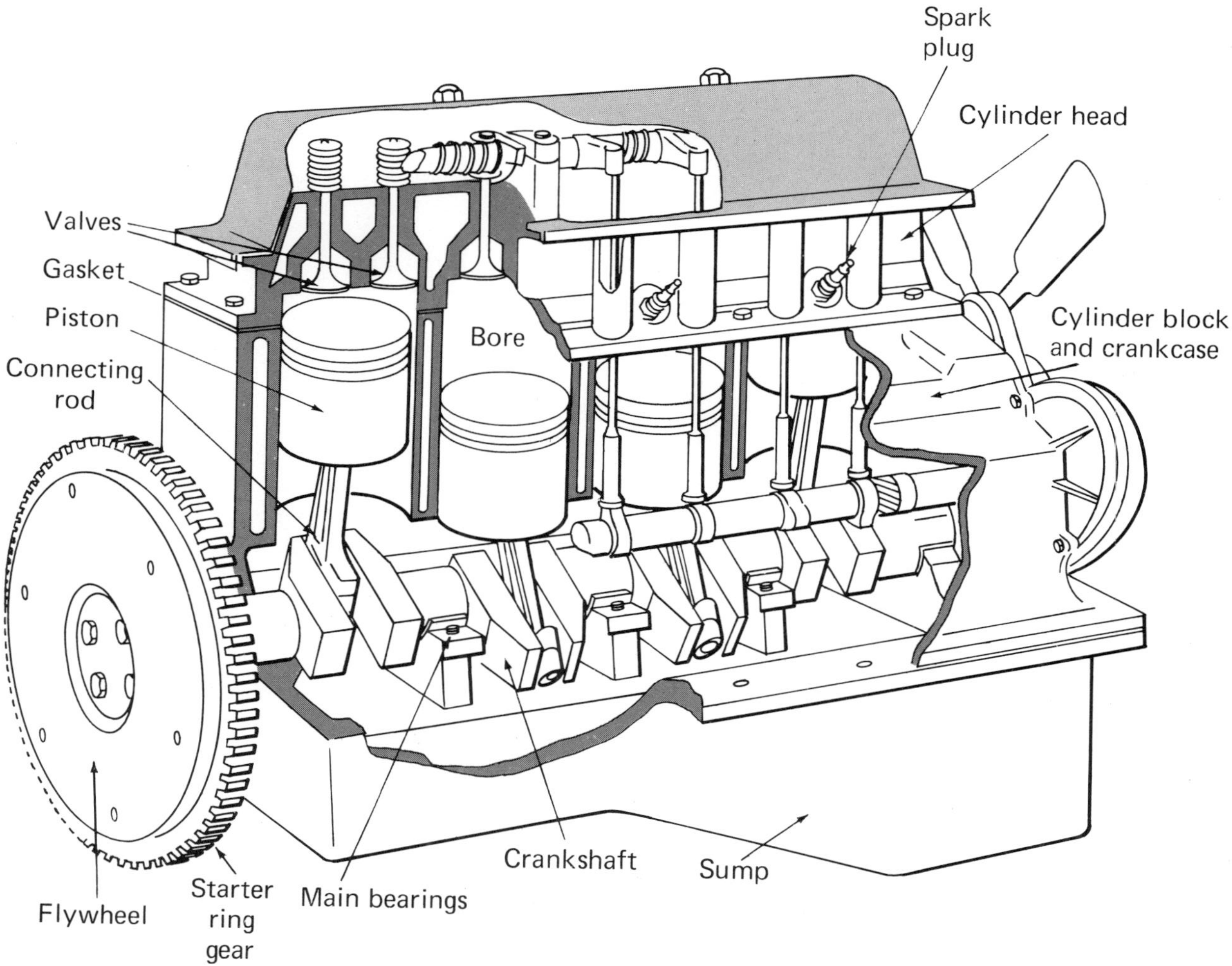
Spark plug
Cylinder head
Valves
Gasket
Piston
Connecting rod
Bore
Cylinder block and crankcase
Flywheel
Starter ring gear
Main bearings
Crankshaft
Sump

*sump.*) The attachment is sealed with a gasket. The part of the block that encloses the crankshaft is the *crankcase.*

Water jackets, which are passageways for cooling water, are in the block around the cylinders. There are passageways and channels in the block for accommodating various parts. The block also serves as a place of attachment for other parts. There are oblong openings in the upper section of the block, to which are fitted the intake and exhaust manifolds. The *manifolds* are systems of pipes that conduct the fuel-air mixture into the cylinders and exhaust gases out. Gaskets are used to seal the lines of attachment.

# operation of the valves

The valves must open and close in synchronization with the movement of the pistons. There are many possible mechanisms for valve operation, but in all systems the motion of the valves is initiated by a rotating *camshaft*, which is turned by the crankshaft. The camshaft is a rod of metal that extends the length of the block. A passageway is bored in the block to accommodate it. The camshaft is driven by the crankshaft either by connection of gears, a chain, or a combination of gears and chain. It has ovoid bumps, or cams, that push against the lifters that are linked to the valve-opening mechanisms.

Almost all the valves used in automobile engines are of the poppet type. They are also called mushroom type because the shape is somewhat like a mushroom. The top or cap of the valve is machined at an angle to fit snugly into the seat of the valve openings. In operation, the valve is moved away from the seat to effect opening.

The valves can be arranged in a variety of positions, either in the head or in the block. The types of arrangements are gener-

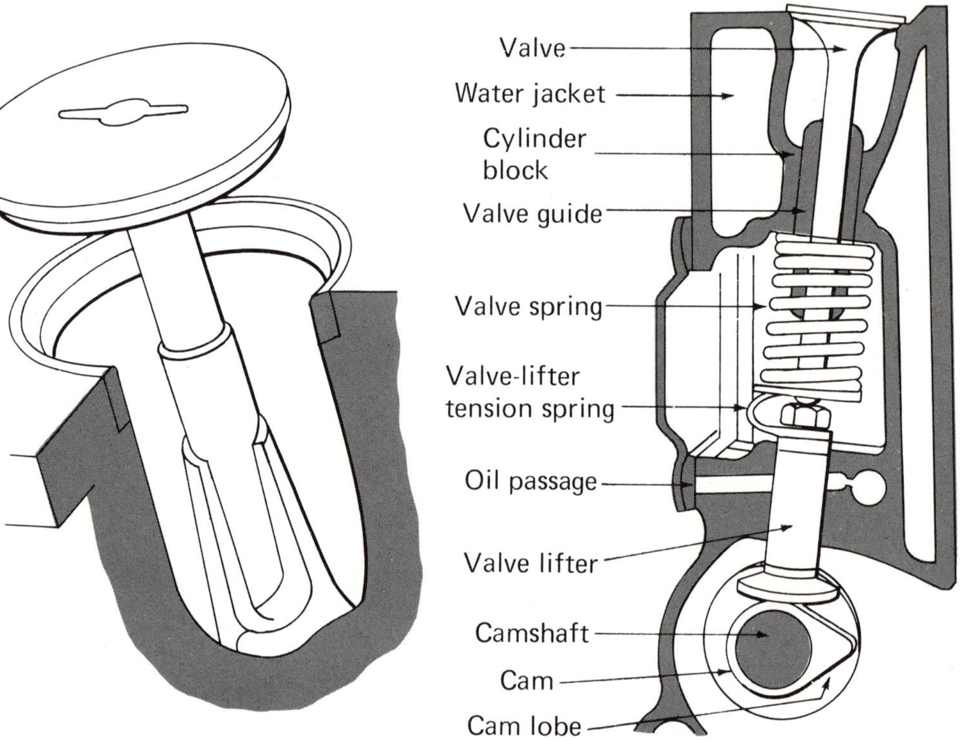


*Left, a valve and valve seat in cylinder; right, typical valve-lifting mechanism*

ally referred to by letters. The usual types in passenger automobile engines are I and T. (The shapes of the cylinder heads suggest the shape of the letters.) L and T heads are also used in some types of engines.

In L-head engines, the valves are placed to the side of the combustion chamber. The valve stems project downward. In overhead valve engines, usually I head, the valves are directly over the cylinders and the valve stems project upward. In recent years, overhead valve engines have become the more prevalent type. The overhead valve arrangement allows for the higher compressions that are desirable in today's high horsepower engines.

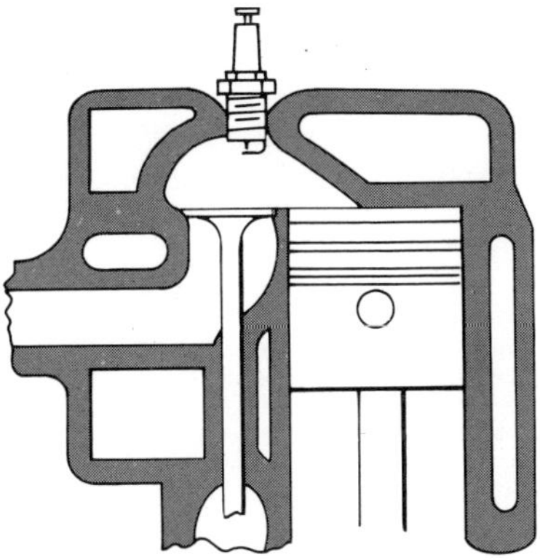

L-Head

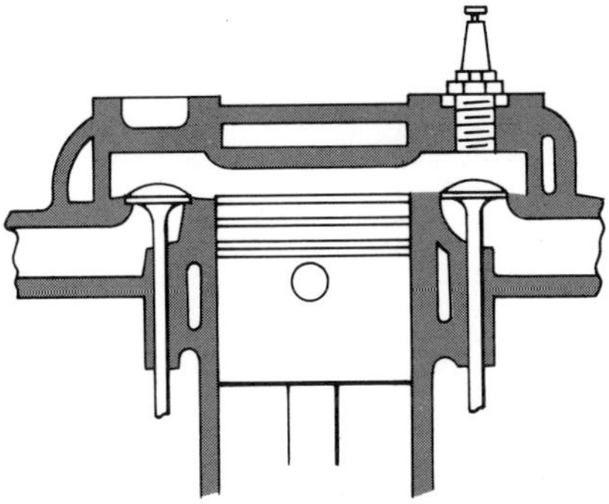

T-Head

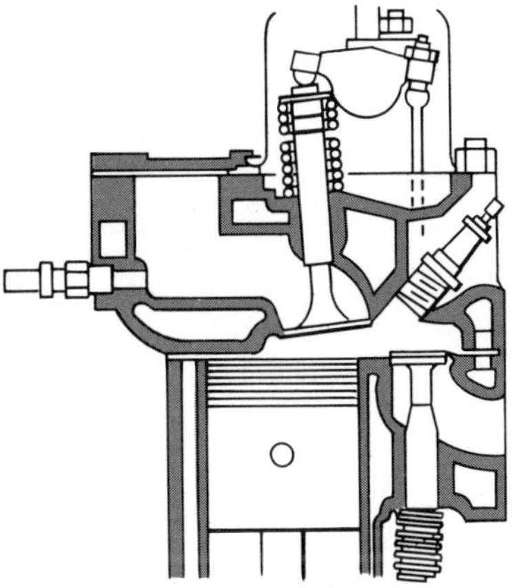

F-Head

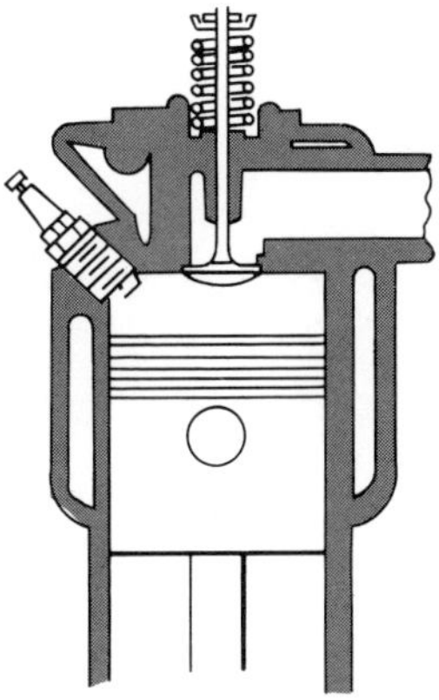

I-Head

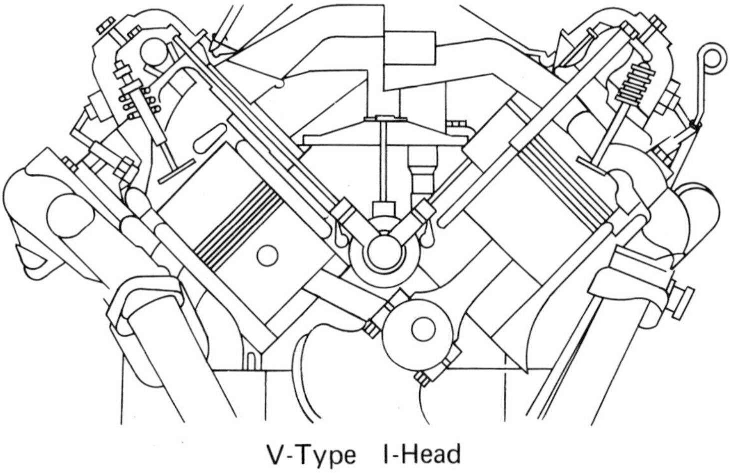

V-Type I-Head

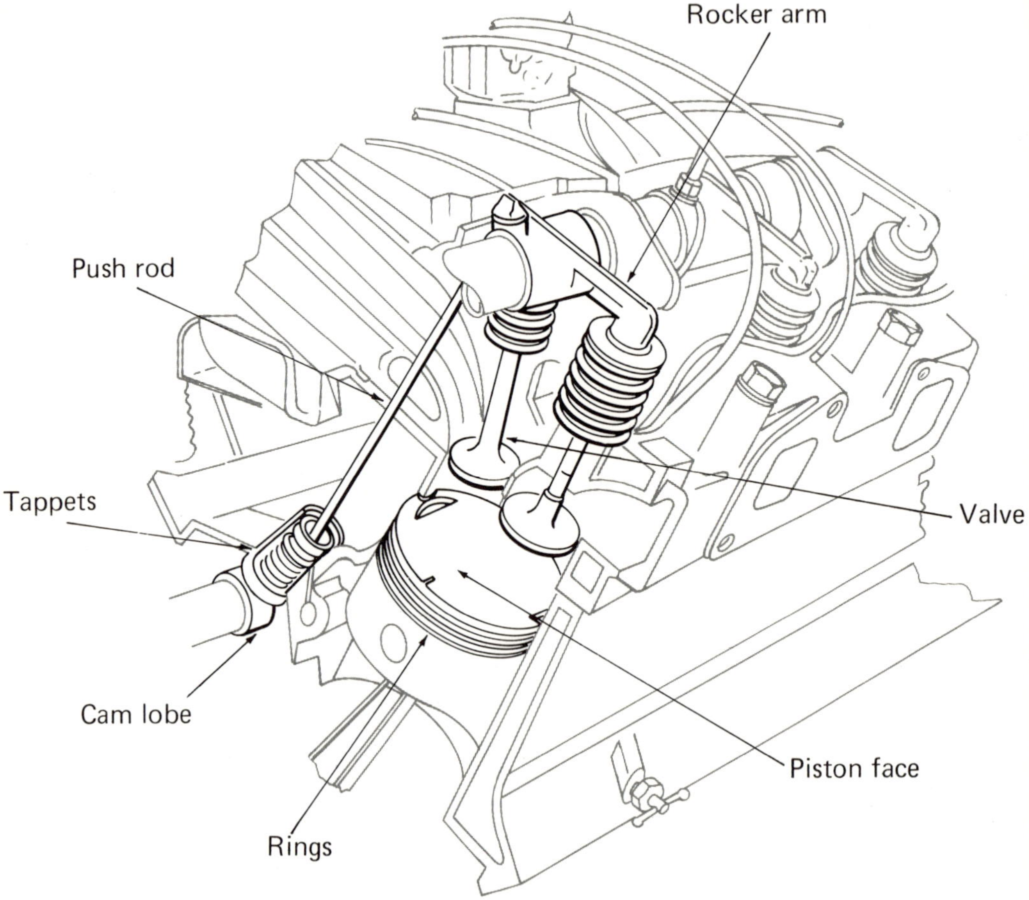


In a typical overhead valve engine, the cam moves against and lifts a *valve lifter*. The lifter is attached to a *push rod*, which is linked with the *rocker-arm assembly*. The valve stem is attached to the rocker arm. The valve-lifter, push-rod, rocker-arm sequence is called the *valve train*. There is a valve train for each valve.

When the push rod pushes against the rocker arm, the other end of the rod pivots down against the valve stem to open the valve. When the rocker arm moves away, a spring closes the valve and holds it shut. Some engines, used in more expensive automobiles, have an overhead camshaft. This eliminates the need for push rods, but requires a more elaborate drive from the crankshaft.

Since the valve train expands as it is heated, a gap or clearance at the valve lifter is required. If there is no gap, the expanded metal in the valve trains will tend to hold the valves open and the valve might be destroyed. The clearances, sometimes called tappets, require frequent adjustments. The hydraulic valve lifter is designed to overcome this problem. It is basically a small hydraulic piston in a cylinder. The hydraulic oil transmits the force to the push rod. Hydraulic lifters have the advantage of being much quieter than mechanical lifters.

# engine characteristics

• *Compression Ratios.* The compression ratio is the comparison of the volume of the cylinder with the piston at the bottom of the stroke to its volume at the high point of the stroke. The relationship is expressed as a numerical ratio, such as 8:1. Higher compression ratios generally imply more efficient engines. Over the years, improvements in engine design have been directed toward achieving higher compression ratios.

The 8:1 ratio is quite prevalent in modern engines. Some racing engines have compression ratios as high as 9.5:1 and even 10:1. Diesel engines, discussed in a later section, have compression ratios around 16:1.

Higher compression ratios require higher quality, more expensive fuels. Compressing the fuel-air mixture causes it to heat up. This might result in combustion before the desired point at maximum compression, causing the so-called engine knocking or pinging. Fuels with high-octane ratings are necessary to retard knocking. Octane rating is a scale of measurement of the fuel's antiknock capabilities. It is based on the burning characteristics of the fuel as compared to that of a substance called isooctane. The antiknock characteristics of isooctane is assigned a value of 100. A fuel with an octane rating of 200 has an antiknock characteristic twice as much as that of isooctane.

Various chemical substances have been added to fuels to increase the octane ratings. The most frequently used has been tetraethyl lead. However, concern over the emission of poisonous lead compounds in the exhaust has forced a marked reduction in the use of lead in gasoline. Automobile manufacturers have been pressured to design engines that will run efficiently on low-lead or no-lead gasolines.

• *Engine Size.* The size of an engine is usually expressed in terms of *cylinder displacement*, which is the volume of the cylinders at piston low point. It is given either in cubic centimeters (cc) or cubic inches. The speed of engines is expressed as *revolutions per minute* (rpm). This refers to the number of revolutions the crankshaft makes in a minute. The power, or work capacity, of the engine is expressed in *horsepower*. A horsepower is defined as the amount of energy required to carry out work at the rate of 550 foot-pounds per second. (Used for measuring work, a foot-pound is the work by a pound of force a distance of one foot in the direction of the force.) Horsepower is measured directly from the crankshaft by a device called a *dynamometer*, which puts a controllable load, or resistance, on the crankshaft. Horsepower so measured is referred to as *brake horsepower*, since the dynamometer exerts a braking force on the crankshaft. This measurement is also called *shaft horsepower*.

The horsepower produced by an engine is a function of such related factors as engine size, rpm, friction, the torque (turning force) applied to and produced by the crankshaft, and many others. Horsepower figures supplied by manufacturers are usually the maximum obtainable at the most favorable rpm. Generally, the horsepower increases with rpm up to a certain point, at which it begins to level off. A higher horsepower engine will not necessarily move the automobile at higher speeds than an engine of lesser horsepower. However, the higher horsepower engine will probably accelerate more quickly and climb hills with greater ease.

# the fuel system

The fuel system of most automobiles consists of a *fuel tank*, a *fuel line*, a *fuel pump*, and a *carburetor*. The fuel tank holds the gasoline, the number of gallons depending on the size of the tank. In early automobiles, the fuel tank was frequently placed directly over the cylinders so that gravity would cause the fuel to drip right into them. Vaporizing was accomplished in a variety of ways. Revolving brushes were tried as well as sponges. But gravity feed frequently did not work on steep hills, and some early automobiles were fitted with hand pumps, which the driver had to operate furiously to get up a hill. Moreover, placing the fuel tank right over the combustion chamber exposed the driver to the dangers of fire and explosions, especially if the engine was mounted in front of the car. It thus became obvious that the end of the automobile, away from the engine, was a far safer place to put the fuel tank. Since tanks are almost always mounted below engine level to allow room for back seats, a fuel pump to move the fuel to the engine is necessary. Rear engine cars have front

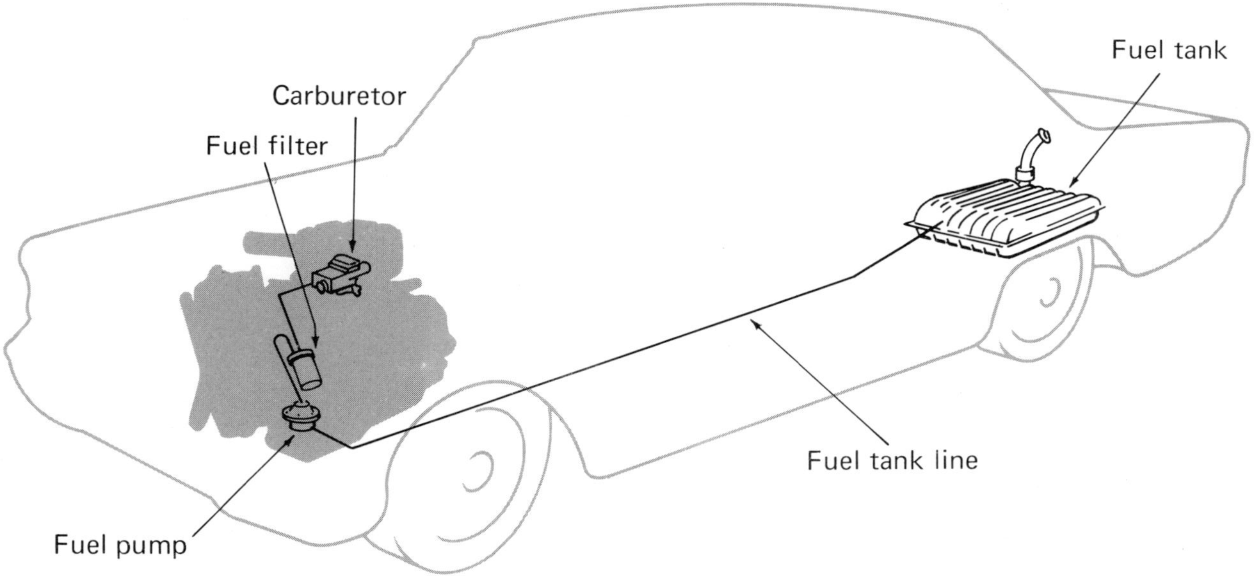
Fuel tank
Carburetor
Fuel filter
Fuel tank line
Fuel pump

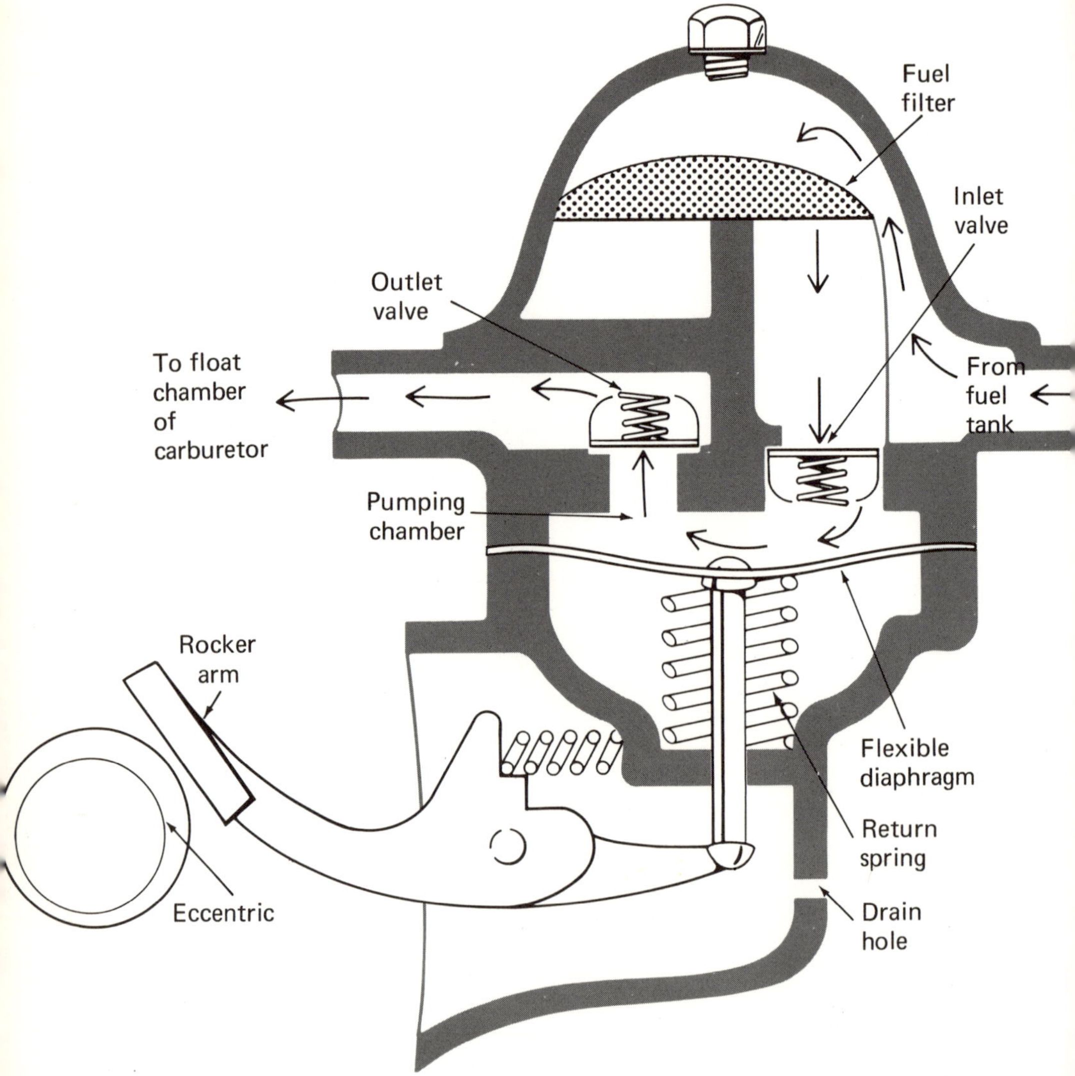


*A mechanically operated fuel pump*

mounted tanks, but a fuel pump is usually installed even if the tank is above engine level. The fuel line is a metal or plastic tube running from the tank to the fuel pump.

Some automobiles have more than one carburetor and others

have a *fuel-injection system* instead of a carburetor. The basic function of the fuel system is to get the fuel from the tank to the carburetor or fuel injector, mix it with the proper amount of air, vaporize it, and feed the proper amount of fuel into the combustion chambers.

• *Fuel Pump.* The fuel pump is placed somewhere along the fuel line between the tank and the engine. Fuel pumps can be mechanical or electric. Since mechanical pumps operate off the camshaft, they must necessarily be mounted on the engine block. (See the illustration on page 25.) Electric pumps can be mounted at any point on the fuel line, but they are most frequently near the fuel tank.

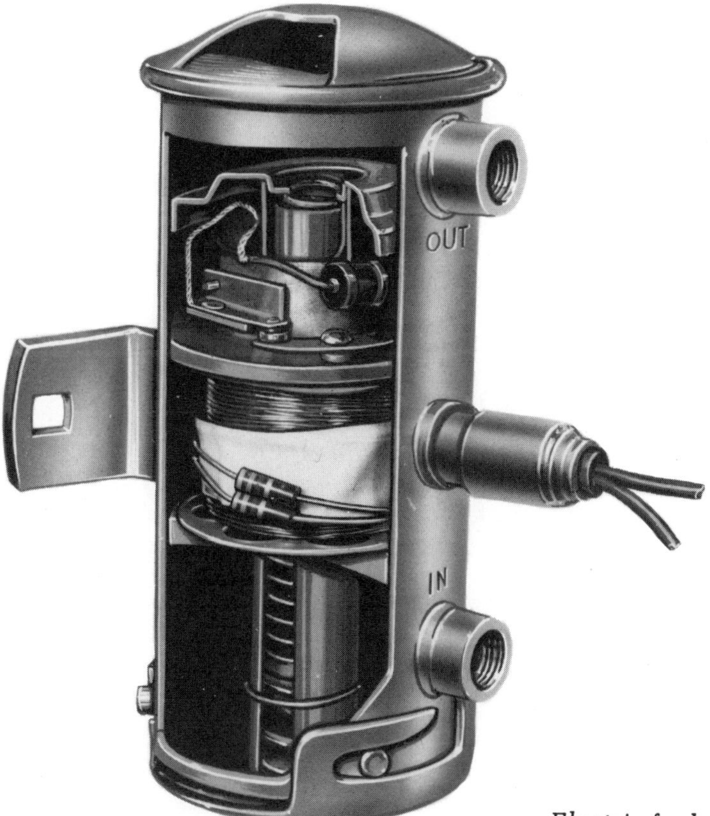


*Electric fuel pump*

In mechanical pump operation, the cam lifts a rocker arm that pulls a flexible diaphragm downward. This brings fuel in by increasing the volume in the pump and, therefore, decreasing the internal air pressure. A return spring pushes the diaphragm up to force fuel out of the pump to the carburetor.

The flow of the fuel out of the pump is controlled by an outlet valve. When the carburetor has, at any given moment, a full load of fuel, the line from the pump to the carburetor is also filled with fuel. The fuel exerts pressure on the diaphragm, preventing it from pumping more fuel into the carburetor.

Electric fuel pumps are found mostly in European automobiles, particularly British cars. The diaphragm operation of the electric fuel pumps is similar to that in a mechanical pump. The diaphragm is attached to a soft iron core that moves when activated by an electromagnet. When the electrical circuit is closed, the electromagnet pulls the diaphragm to bring fuel into the pump. When the circuit is broken, a return spring pushes the diaphragm back to force fuel out of the pump and into the carburetor. A rod, usually made of bronze, moves with the iron core to activate a flick over mechanism that separates a set of contact points, which breaks the circuit. The bronze rod also moves with the return spring to close the contact points. The speed of the pump operation is controlled by the amount of current received from the automobile's electrical system.

• *Carburetor.* The purpose of the carburetor is to mix the fuel with air, vaporize the mixture, and deliver the fuel-air mixture to the cylinders. The carburetor is designed to carry out its function in all operating conditions — low speed, high speed, idling, accelerating, decelerating, and so on.

The carburetor in its simplest form is the downdraft carburetor. The flow of the mixture is aided by gravity. Fuel from the fuel pump is fed into the *float chamber.* As fuel comes in, the float rises. At a certain point, a *needle valve* on the float plugs up the tube leading from the fuel pump. As the level of fuel drops, the float drops and more fuel can come into the chamber.

Air flows downward into the opening, or throat. The opening is covered by an air cleaner that filters out dirt particles into the barrel of the carburetor. The carburetor narrows at a point called the *venturi* to increase the speed of the air rushing down through the carburetor. The downward movement of the pistons in the in-

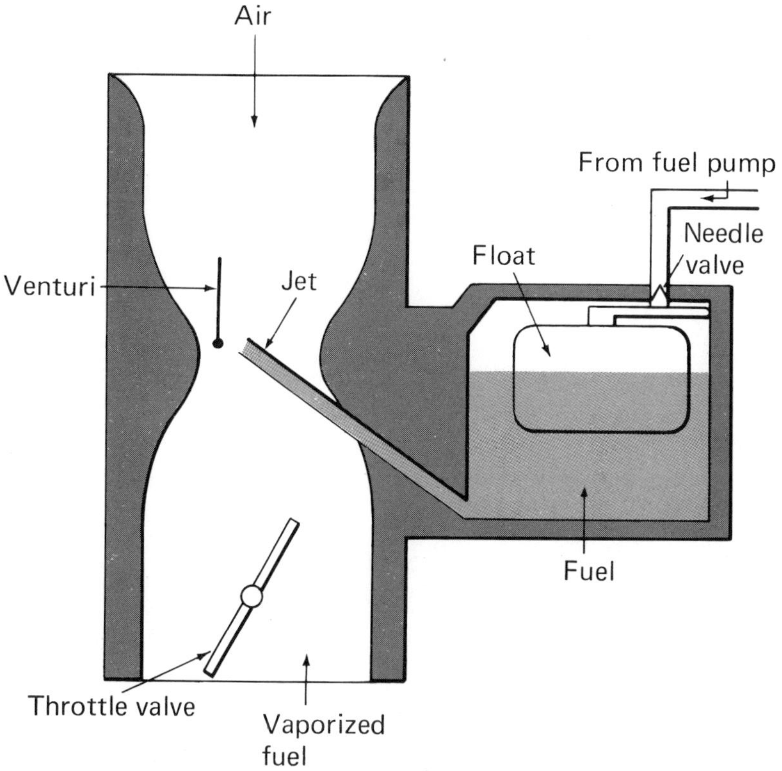


*Diagram of a simple carburetor*

take valves sucks air down through the barrel of the carburetor and speeds up the flow of air. The speeding up of the passage of air into the pistons causes a decrease in atmospheric pressure in the venturi. This decrease in air pressure allows fuel to flow into the air stream from the *jet* that leads from the float chamber. The fuel and air are then mixed and vaporized.

The amount of gasoline-air mixture that enters the intake manifolds and then moves to the cylinders is controlled by the *throttle valve*, generally referred to as the throttle. It is a disk of metal that turns on a spindle. It is controlled by the *accelerator*, or gas pedal. When the accelerator pedal is pressed, a linkage causes the throttle to open to allow more mixture into the intake manifolds.

The amount of air entering the barrel is controlled by the *choke valve* (or choke), which is similar in structure and operation to the throttle. It is also mounted on a spindle. The choke is generally closed when starting a cold engine. This allows less air to enter the combustion chamber, thus creating a "richer" (more gasoline per amount of air) mixture, and makes ignition easier. Some automobiles have chokes controlled by the driver through a knob on the dashboard, connected to a linkage. Most modern automobiles, however, have automatic chokes that are opened and closed by a thermostat.

As the engine runs faster, the air flow increases, and this increases the flow of fuel. The ratio of air to fuel varies with the positions of the choke and throttle valves. An ideal mixture for cruising is about 15 parts of air to 1 part of fuel (by weight). If the amount of air is greater than this, the mixture is lean. If the ratio of air to fuel is less than 15:1, the mixture is rich. Different degrees of leanness and richness are needed for best operation in different driving conditions.

Most carburetors are more complex than the simple downdraft type. The simple carburetor cannot vary the mixture to suit different conditions. At high speeds, the mixture would tend to be too rich, and at low speeds, the mixture would be too lean. This is because air and liquid flow at different rates. To function well in high-speed, powerful engines, carburetors must be equipped with devices to compensate for the different flow rates of air and fuel.

A *compensating jet* and an *air blend* provide for different routes of fuel flow for idling, low speeds, and high speeds. In idling, the fuel flows from the float chamber, past an idle-adjusting screw, to enter the barrel just below the throttle. At idle, the throttle valve almost completely closes off the barrel. The vacuum that is formed on the manifold side of the throttle valve activates the idle flow mechanism. As the name implies, the idle-adjusting screw can be turned with a screwdriver to effect faster or slower idling.

At higher speeds, the throttle is open. The vacuum below the throttle valve is reduced so the fuel flows through the main jet. The air bleeder, or high-speed bleeder, admits more air to prevent the mixture from getting too rich.

Carburetors can have two or four barrels. This improves the distribution of fuel over the wide range of speeds. In a two-barrel

*Four-barrel carburetor in V-8 engine*

carburetor, each barrel supplies half the cylinders. In some four-barrel carburetors, two barrels are in use at low speeds and all four at high speeds. Some types of racing and drag vehicles have what approaches the ultimate in carburetors. Each cylinder has its own. Such engines require constant adjustment for most efficient operation.

• *Fuel Injection.* In recent years, fuel injection, which has always been used in diesel engines, has been made available for gasoline engines. Fuel injection is much more efficient than carburetion, but it is also more expensive and is usually offered as an option at extra cost.

Fuel-injection systems squirt precisely measured amounts of fuel into the cylinders at great pressure, generally at 200 psi (pounds per square inch). At the heart of most fuel-injection systems is a complicated pump mechanism. The pump forces the fuel at high pressure to a nozzle at each cylinder. In most gasoline engines, the nozzles are located in a position to squirt the fuel into an intake port near each cylinder. The fuel is mixed with air at the intake port.

Fuel-injection systems have control mechanisms that compensate for manifold pressure, engine speed, and the temperature of the air in the intake manifolds. The control mechanisms are mechanical, thermostatic, and electronic.

Fuel injection overcomes many of the problems that are frequently encountered with carburetors. It distributes the fuel more equally, and this improves economy. Carburetor "flooding," in which an excess of nonvaporized gasoline gathers in the carburetor, does not happen with fuel injection. When the carburetors are flooded, starting the car is very difficult. Since the fuel is distributed equally with fuel injection, such an automobile would tend to have more pep and power than the same car with carburetion.

# the electrical system

Electricity is needed to provide the spark for igniting the fuel in the cylinders, for starting the car, and for operating all the electrical devices in an automobile — lights, horn, radio, heater, air conditioner, and others. Almost all modern automobile electrical systems are the battery-coil type. The main function of the electrical system is to provide current for producing a spark to ignite the fuel. The main parts of the ignition system are the *battery*, the *coil*, *points* (housed in the distributor casing), the *distributor*, and *spark plugs*. The *generator* and *voltage regulator* may also be considered part of the ignition system.

• *Battery*. Most modern automobiles have twelve-volt batteries. The battery produces electricity by chemical action. In well over 90 percent of today's batteries, the electricity is produced by the action of sulfuric acid on compounds of lead. Each cell in the battery produces two volts of electricity. The voltage of a battery can be determined by counting the ports and multiplying that number by two. Each port represents a cell, to which water is

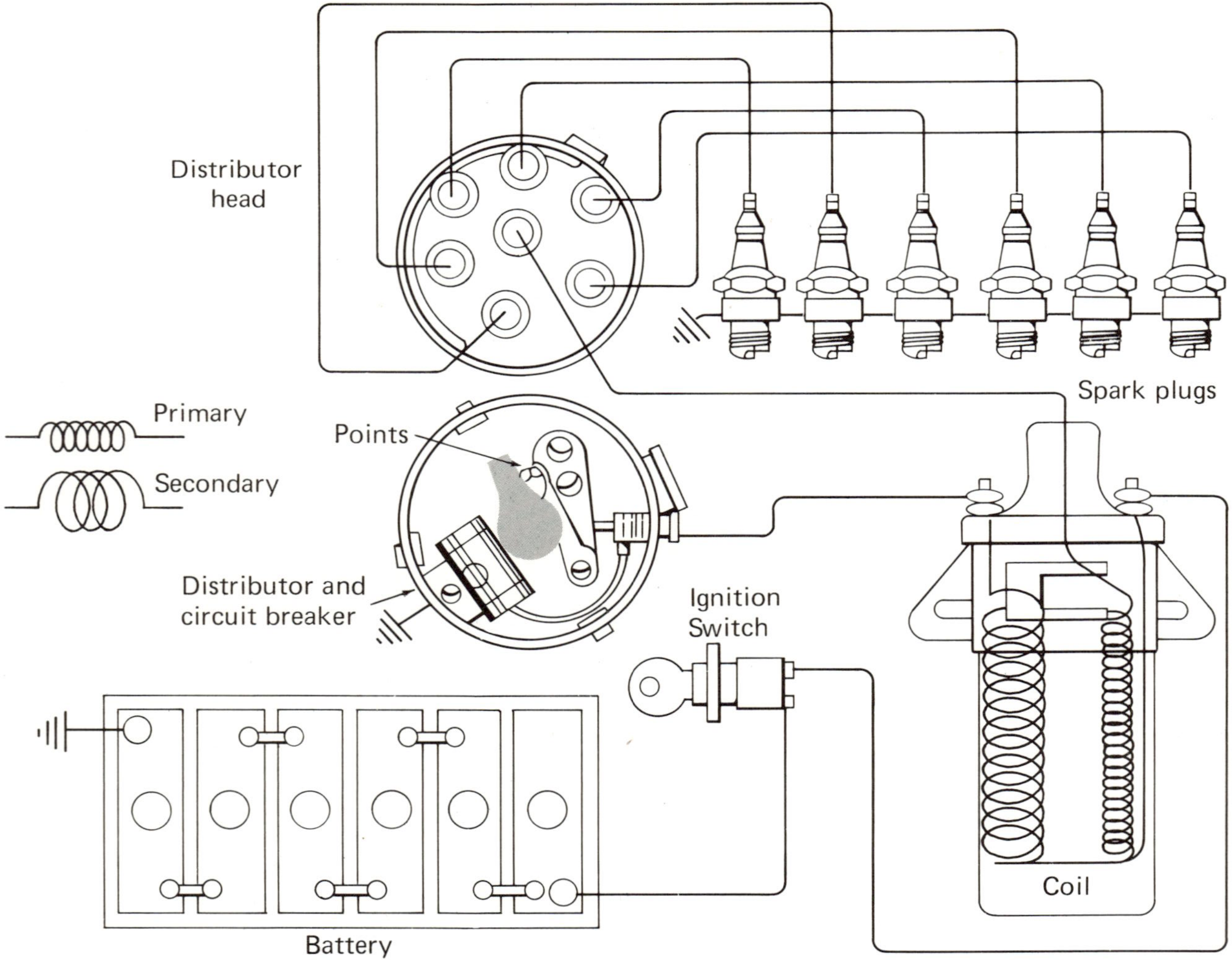
Distributor
head
Spark plugs
Primary
Secondary
Points
Distributor and
circuit breaker
Ignition
Switch
Coil
Battery

added. The sulfuric acid in the battery is diluted with water. In the normal course of operation, some of this water is lost and has to be replaced from time to time. The battery has two leads, or posts, marked + and −, the anode and cathode, respectively. The cathode is grounded to the engine block and ultimately to the body of the car via a cable. Electricity flows from the anode through a cable to the starter and other electrical parts. The circuit is closed when the ignition key is turned on.

- *Ignition Coil.* Voltage is electrical pressure or potential. It can be compared to the pressure of water in a pipe or hose. The twelve volts of a car battery are not enough to push the current through the spark-plug gap. The gap is air that has a high resistance to electric current. The voltage is increased from twelve to about twenty thousand volts by the ignition-coil system.

The current goes from the battery through a set of contact points (commonly just called "points") into the primary wire winding of the coil, the secondary wire winding of the coil, and then to the distributor. The coil is essentially an electrical device called a step-up transformer.

When the contact points are closed, low voltage current from the battery flows into the primary winding. This sets up a magnetic field around the windings. When the points open, the circuit is broken and the current flowing through the primary coil suddenly collapses to zero. This induces a high-voltage current in the secondary coil winding, and this current goes to the distributor.

The points are opened by a tension spring and closed by a cam on a rotating shaft. The rotating shaft is driven by the camshaft. Sparking can occur when the current jumps the gap between the open breaker points. This can burn and pit the points. But a *condenser,* or capacitor, placed in the point circuit draws off the electricity when the points are open. The condenser helps to effect a clean break between the points, which in turn aids the building of high voltage in the secondary coil.

*Main parts of the ignition system*

• *Distributor.* As the name implies, the distributor distributes current to the spark plugs. Basically, the distributor is a rotating switch with as many contacts as the engine has cylinders. High-voltage current from the ignition coil enters the distributor and passes into the rotor. The rotor is placed on top of a rotating shaft that is driven by the camshaft. The rotation of the shaft is timed either through a set of gears, a chain, or a combination of chain and gears. The rotating distributor shaft also bears the cam that closes the breaker points. Contacts are on the undersurface of the *distributor cap*, which is placed over the body of the distributor like an inverted cup. As the rotor switch goes around, it makes and breaks contacts, distributing current to the spark plugs in a predetermined sequence. Since the rotor is turned ultimately by the engine, it increases speed as the engine increases speed. Since the breaker points are controlled by the same shaft as the rotor, the operation of the points is in phase with the distributor.

• *Spark Plugs.* Cables extend from each of the distributor contacts to a spark plug. They are pushed onto the spark-plug terminals and can be pulled off easily when the plugs have to be changed or cleaned. A spark plug is a metal electrode that carries the current from the wire that is connected to the distributor. The electrode is surrounded by a porcelain insulator. Below the insulator is a threaded metal portion that screws into the cylinder head. The business end of the spark plug, which is the threaded metal part, is the gap between the electrode and the ground. When electric current jumps the gap between the electrode and the ground, a spark is produced. The size of the gap varies, but is generally between .015 inch and .035 inch. Automobile manufacturers recommend the best gap sizes for the engines they produce.

Spark plugs become corroded with use and must be replaced from time to time. How often is dependent on many things, not the least of which is how fussy the owner is about his car's performance. As plugs wear and corrode, the automobile's performance and fuel economy are affected. When plugs are replaced, it is usually advisable also to replace the breaker points and condenser, and to adjust the timing. This is what is known as a "tune-up."

• *Spark Advance.* The fuel in the cylinders burns very quickly, but there is a passage of time, measured in fractions of a

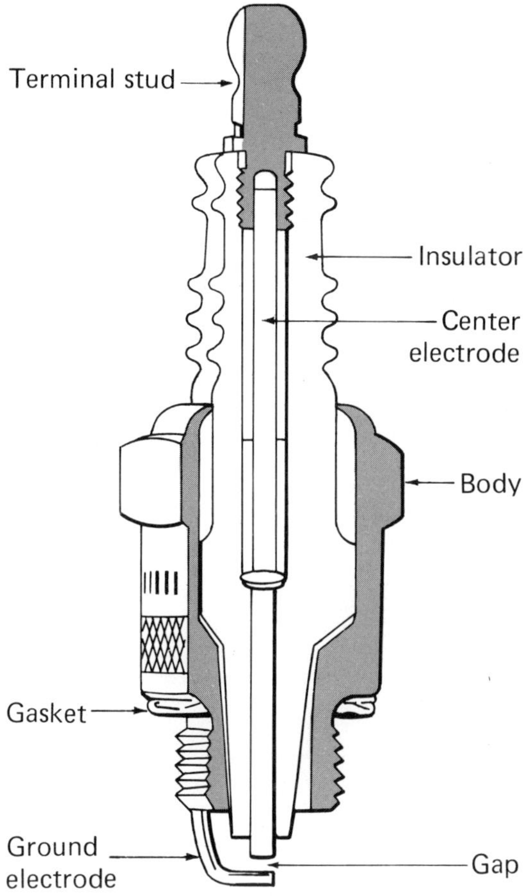


*Diagram of a spark plug*

second, before the fuel burns sufficiently to supply expansive power. This could result in rather erratic engine operation or no operation at all. A spark-advance mechanism causes the spark to occur just before the piston has come up to its highest point of travel ("top dead center" is the term used to designate this highest point) at slow speeds, and quite a bit before top dead center at high speeds. This enables the expansion of gases to occur just as the piston starts on its downward stroke. An advance mechanism is required to vary the spark timing as needed. Advance mechanisms operate off centrifugal weights and vacuum devices that utilize intake manifold vacuum.

• *Generator.* The battery can produce current for only a limited time. Eventually, it runs down and it must be recharged. Actually, the battery is continuously being recharged by the generator, which is run by the engine via a belt and pulley. The generator operates on the principle that when a coil of wire, such as copper wire, is moved in a magnetic field, a movement of electrons — an electric current — is produced in the wire. In a generator, many coils of wire are wound on an iron core that is attached to a rotat-

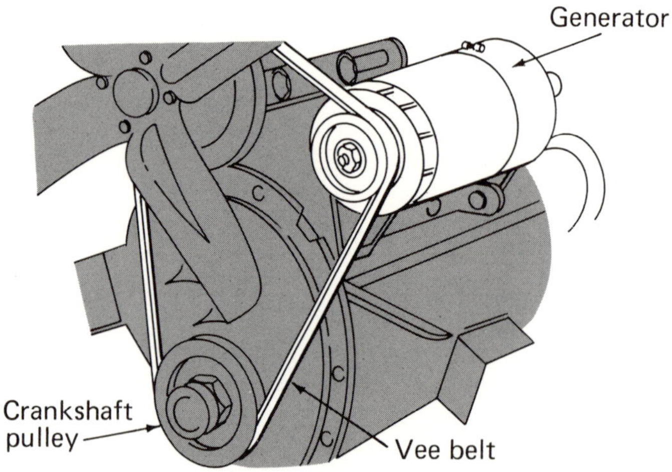


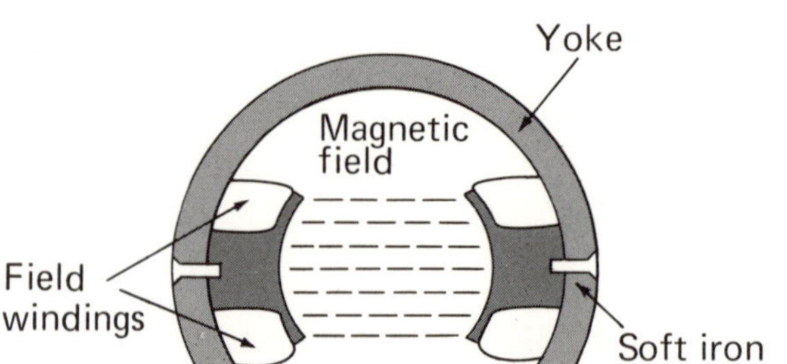


Cross-section of armature

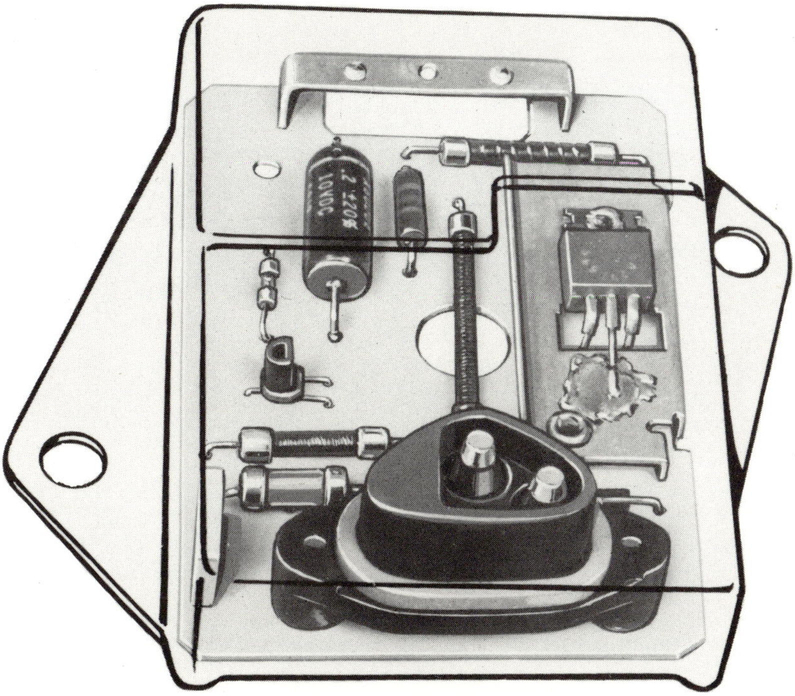

*Electronic voltage regulator*

ing shaft. This is called the *armature*. The magnets may be permanent or electromagnets and are situated against the body or casing of the generator assembly. As the armature rotates, current is fed to the battery as it is needed.

The current is picked up by carbon contacts called *brushes*. The brushes are in direct contact with the armature, so they eventually wear down and must be replaced. Alternating current produced by the generator is changed to direct current by a *commutator*. Only direct current can recharge the battery. The current from the generator effects chemical changes that are the reverse of the chemical changes that produce current in the battery.

If the generator kept feeding current to the battery after the battery was sufficiently charged, the battery would be destroyed. A *voltage regulator* and *cutout device* controls the amount of current fed to the battery. The cutout is an automatic switch in a

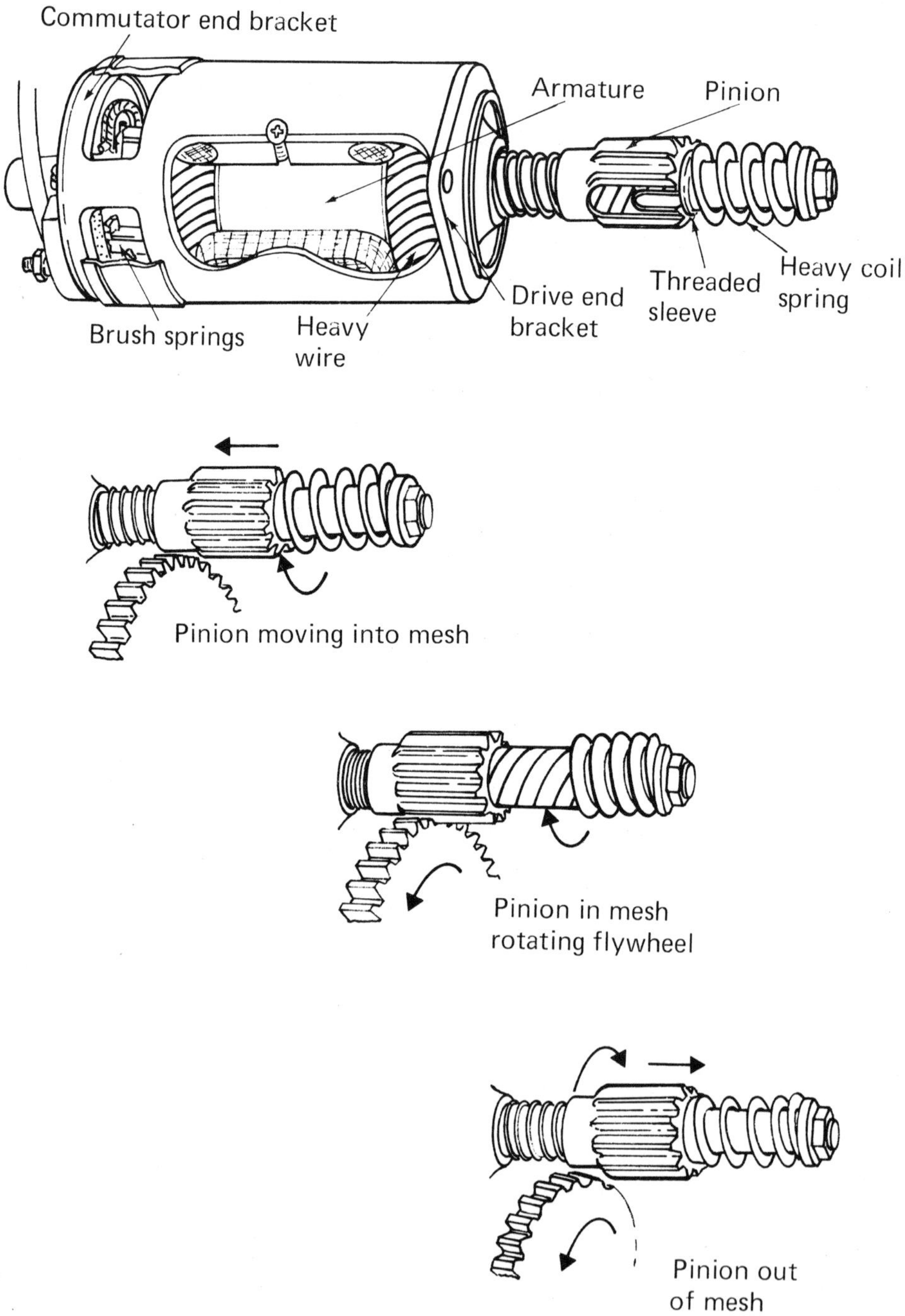

Commutator end bracket
Armature
Pinion
Threaded sleeve
Heavy coil spring
Drive end bracket
Brush springs
Heavy wire
Pinion moving into mesh
Pinion in mesh rotating flywheel
Pinion out of mesh

circuit between the battery and the generator. The contact points in the cutout are opened and closed by an electromagnet.

The voltage regulator is somewhat misnamed, since it consists of voltage and current regulator units. Both are operated by electromagnetic fields. The current regulator allows maximum current to flow into the battery. When the battery approaches full charge, the voltage regulator is activated and it maintains a near-constant voltage in the battery.

• *Alternator.* Alternators are a different type of generator. They work on the same basic principle of electromagnetic induction as other automotive generators. Electromagnets rather than permanent magnets are used to produce the magnetic field. The strength of the magnetic field can be varied to meet the demands imposed by the many electrically operated parts of a modern automobile.

In the alternator the magnet rotates. In most other types of automotive generators, the armature rotates. The magnet rotates in a case that has three separate windings and generates a three-phase alternating current. The current is rectified, that is, changed to direct current, through a system of *diodes* or *rectifiers*.

• *Starter.* Internal-combustion engines require some outside force to get them started. In early automobiles, the outside force was the driver or an accommodating friend. A crank handle was inserted into a nut on the front of the crankshaft. The crank handle turned the crankshaft and, hopefully, the engine started. The engine of the 1875 Marcus vehicle was geared directly to the rear wheels. Starting the Marcus required a strong man to lift the back end so the rear wheels could be spun by hand.

Modern starters are electric motors mounted at the flywheel. The flywheel is circled by a gear called the *ring gear.* A shaft from the starter motor terminates in a small *pinion gear.* When the starter motor is turned on, the pinion gear meshes with the ring gear and turns the crankshaft.

As soon as the engine starts, the pinion gear disengages from the ring gear. This is accomplished by the pinion spinning along a

*The starter motor*

threaded shaft. The action takes place as soon as the engine is turning at a faster rate than the starter motor.

In most cases, the starter switch circuit is built into the ignition switch assembly. Giving the key an extra turn activates a type of electromagnet called a *solenoid,* which closes a circuit to send current into the starter motor and to engage the ring and pinion.

# lubrication

As a car engine runs, many metal parts rub against each other. The resulting friction will rapidly wear out the engine parts unless a film of a slippery substance such as oil is introduced between the moving parts.

When the layer of oil separating the moving parts is so thin that it is measured in thousandths of an inch, the lubrication is referred to as *boundary lubrication. Fluid lubrication* occurs when the parts are practically bathed in oil. Only the main bearings of the crankshaft approach the fluid lubrication condition. Most other parts are boundary lubricated.

Early cars depended on the *splash system*. The crankshaft was in a pan of oil and its turning (with luck) splashed oil over the rest of the moving parts of the engine. In modern engines, the oil is circulated to the various moving parts through lines at high pressure. Some parts of modern cars, such as the cylinder walls, pistons, small ends, and piston pins, are still splash lubricated. The oil also helps to cool the engine.

Oil is stored in the sump, or oil pan, which is fitted to the

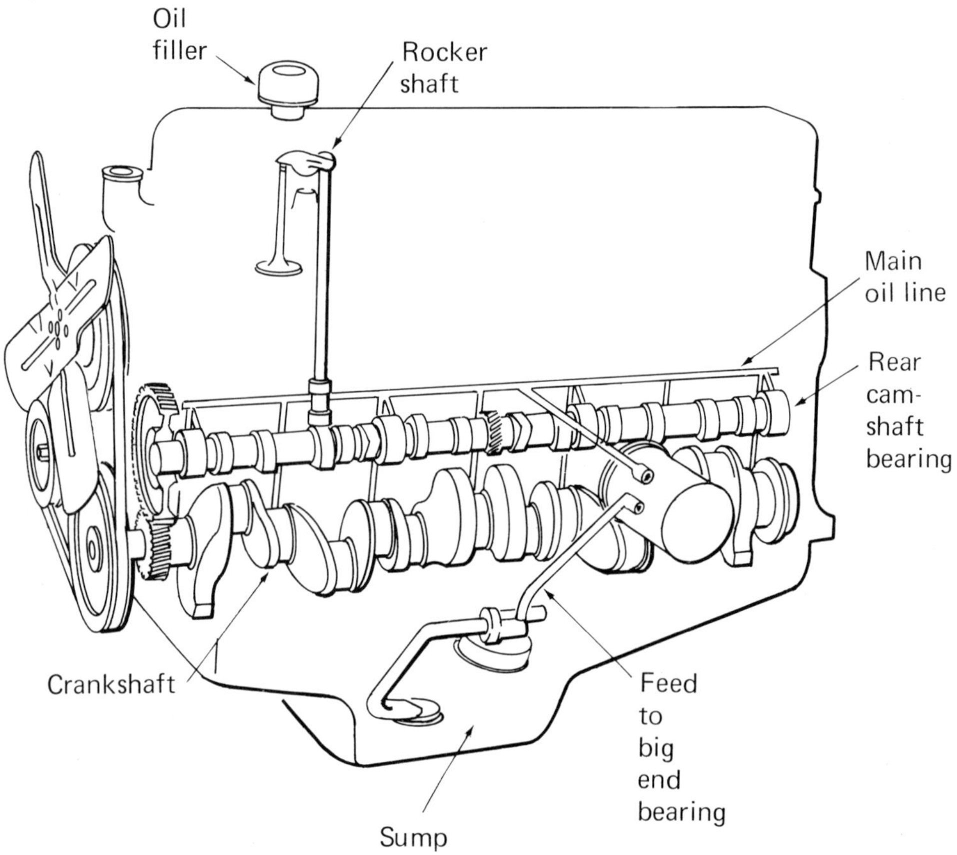


*Pressure lubrication system*

bottom of the block. An *oil pump* is immersed in the oil. The pump is driven by the camshaft. There are various types of oil pumps. The *gear type* is most frequently used. A driving gear, running from the crankshaft, turns another gear (the driven gear) of similar size. The gears are mounted in the pump housing. The motion of the gears moves the oil upward along the outside gear teeth somewhat like paddles.

The oil passes through a coarse filter immersed in the sump on its way to the pump, and through another filter after it leaves the pump. In a typical engine, oil flows from the filter to a main oil passage that runs the length of the engine. A *pressure relief valve* prevents engine damage from oil pressure that is too high. The valve is a ball on a tension spring. When the pressure is too high, the spring is pushed against the ball, which opens a port and allows oil to flow back to the sump. The relief valve may be in the pump or in the main oil passage. Oil from the top of the engine drains back to the sump through the push-rod passages, lubricating them and the valve lifters on the way.

Oil filters fill up with debris and should be replaced frequently. The oil itself gets dirty, and should be drained and replaced with new oil at regular intervals. Most oil filters are full flow. That is, all the oil from the pump flows through them before going to the main passage. The filter is a metal can with filtering material, such as felt or paper, in it. Most have a bypass to allow flow to continue if the filter elements become clogged.

Oils are available in a variety of viscosities (degrees of thickness) for different conditions. The Society of Automotive Engineers (SAE) has devised a scale of 5 to 50 to designate oil viscosity. A lower number indicates a lower viscosity. Since oil gets thicker when cold and more runny in heat, thicker, more viscous oils are generally used in summer, and lighter oils in winter. Some oils have multiratings, such as 10–30. They contain additives that keep the oil from getting too thick at winter temperatures, but they do not thin out at summer temperatures. Oils may also have other additives such as detergents (to break up gummy deposits), corrosion retardants, and oxidation retardants.

# cooling systems

The rapid combustion in automobile engines produces high heat, which would soon destroy the engine if it were not removed by a cooling system. Cooling systems are of two general types — air and water.

• *Air cooling.* This type of cooling is used mostly in aircraft, motorcycles, and gasoline lawn mowers. However, it is also used in the Volkswagen, one of the largest-selling automobiles in the world.

Air cooling works by means of fins on the cylinders that increase the surface area exposed to the air. A fan blowing over the fins and the motion of the car remove heat.

Air-cooled engines warm up faster than water-cooled engines, and since there is no water to freeze in winter, no antifreeze is needed. However, if they do overheat, these engines can be quickly damaged. If water-cooled engines overheat, all the water must boil off before serious damage occurs. The overheating is usually noticed before this happens. Air-cooled engines tend to be noisy, since there is no water around the engine to act as a sound absorber. Also, the efficiency of the air-cooled system tends to go down as engine size goes up.

• *Water cooling.* Water has a very high specific heat — a given volume of water will remove more heat than the same volume of most other natural substances. In water-cooled engines, water circulates in a *water jacket* around the cylinders. The heated water moves to the *radiator*, where heat is removed by air blowing over the water. The air flow comes from the motion of the car and from a fan, which, in most cars, is driven by a *fan belt* and pulley on the front of the crankshaft. In most cars, the same belt that drives the generator also drives the fan.

The radiator is a narrow tank with many passages that divide the flow passage of the water. The purpose is to expose the largest possible surface area to the air.

The water-cooling system in its simplest form is the thermosiphon system. As the water in the water jacket is heated, it tends to rise, due to convection, up to an outlet hose mounted on top of the block. The outlet hose leads to the top of the radiator. As the water is cooled in the radiator, it gets heavier and moves downward to the lower tank of the radiator. An inlet hose leads from the lower hose to the water jacket, and the cycle is repeated.

Although the thermosiphon system is still used in some small cars, the action is too slow for large, powerful engines. And the necessity of mounting the radiator above engine level affects the design of the car. So, most cooling systems have a water pump.

The water pump is a reasonably simple mechanism and is driven by the same belt and pulley system that drives the fan. In fact, it is usually on the same spindle as the fan. Convection currents are still at work, but the pump helps to move the water from the lower tank to the water jacket.

Water-cooling systems are usually equipped with a thermostat that controls the rate of water heating and cooling. The thermostat provides for a quicker warm-up by blocking the flow of water to the radiator during the warm-up period. Usually placed at the cylinder-head outlet, the thermostat is generally a flexible metal bellows filled with a volatile liquid. A flat valve is attached to the bellows. At cold temperatures, the contracted bellows holds the valve against the entrance to the overhead

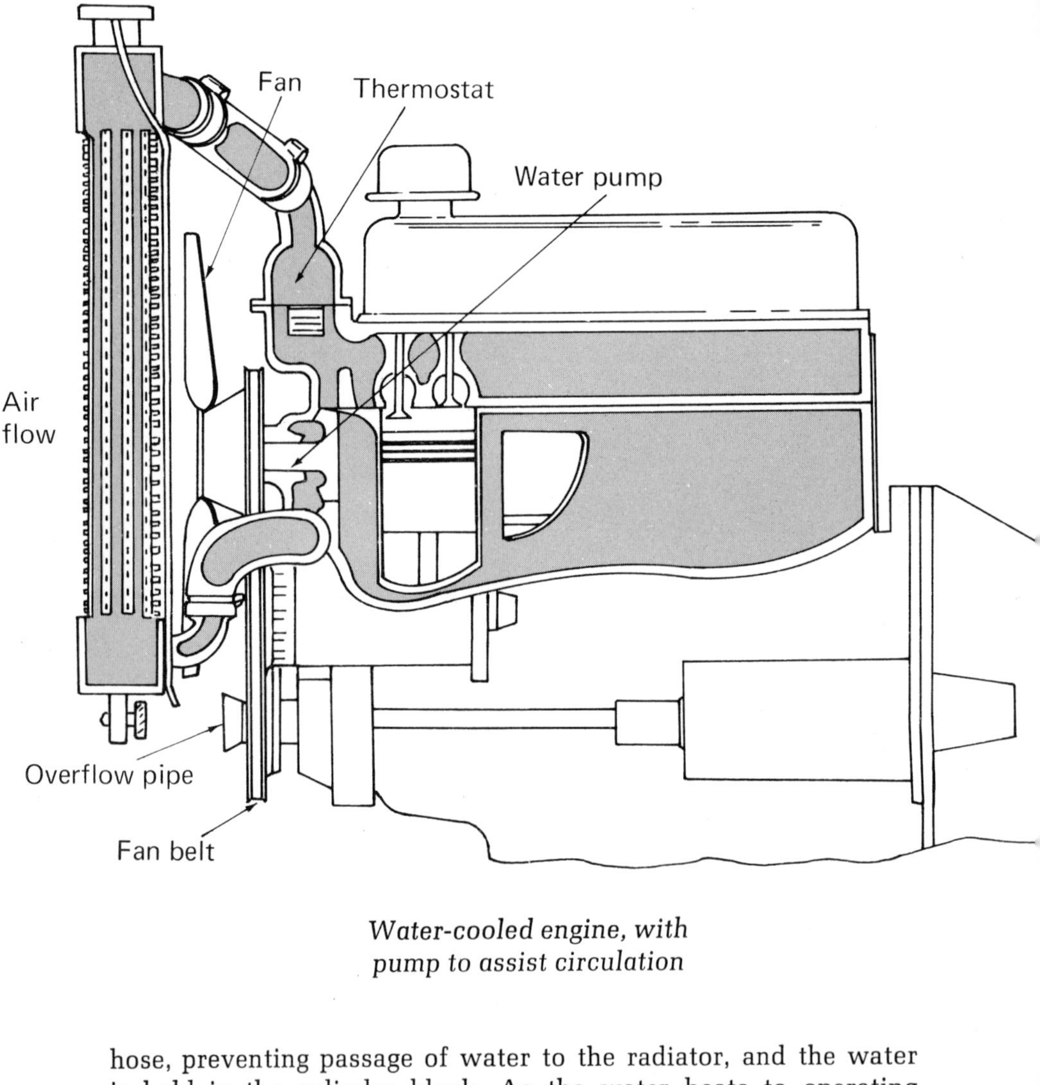


*Water-cooled engine, with pump to assist circulation*

hose, preventing passage of water to the radiator, and the water is held in the cylinder block. As the water heats to operating temperature (about 180 degrees F), the liquid in the bellows boils, causing it to expand. This opens the valve, and water flows into the radiator.

Water is added through an opening at the top of the radiator. An overflow pipe directs excess water out of the radiator to the ground. In most modern automobiles, the opening is closed with a *pressure cap*. When in place, the cap seals off the overflow pipe to prevent loss of water. The pressure cap has a valve and spring that allows high atmospheric pressures in the cooling system. As the atmospheric pressure goes up, the boiling point of water increases, so the engine can operate at temperatures above the boiling point of water, and the water will not boil away. When the engine cools, the spring valve contracts to allow the pressure to drop to atmospheric levels. This prevents damage to the hoses and other fittings.

Antifreeze, usually ethylene glycol, is added to the water to lower the freezing point of the water. Water has the peculiar property of expanding after it freezes. Freezing of the cooling water can crack the block. Some cars today are built with permanently sealed cooling systems filled with a low-freezing-point liquid. As long as no leaks develop, water never needs to be added.

# exhaust systems and emission controls

The exhaust system is a series of pipes that carry exhaust gases to the outside of the engine. The system also deadens engine noise. In most cars, the exhaust system consists of the *exhaust manifold,* the *exhaust pipe,* the *muffler,* and the *tail pipe.* The exhaust manifold is a complex of metal passageways attached to the cylinder block that collect the exhaust gases. The manifold passages merge into the exhaust pipe.

In some engines, especially V-8s, there is a dual exhaust system. The manifolds from each side merge into an exhaust pipe. Two separate manifolds and pipes reduce back pressure, which is a force that acts on the piston. One exhaust pipe requires an extra "bend" around the engine. This increases back pressure. Back pressures tend to resist the upward movement of the piston in exhaust stroke, reducing the power output of the engine. Some racing cars have exhaust pipes leading from each cylinder.

Since the motive force of the engine is supplied by rapidly burning fuel, the noise produced would be ear-shattering if it were not muffled. Mufflers are metal casings, usually equipped with

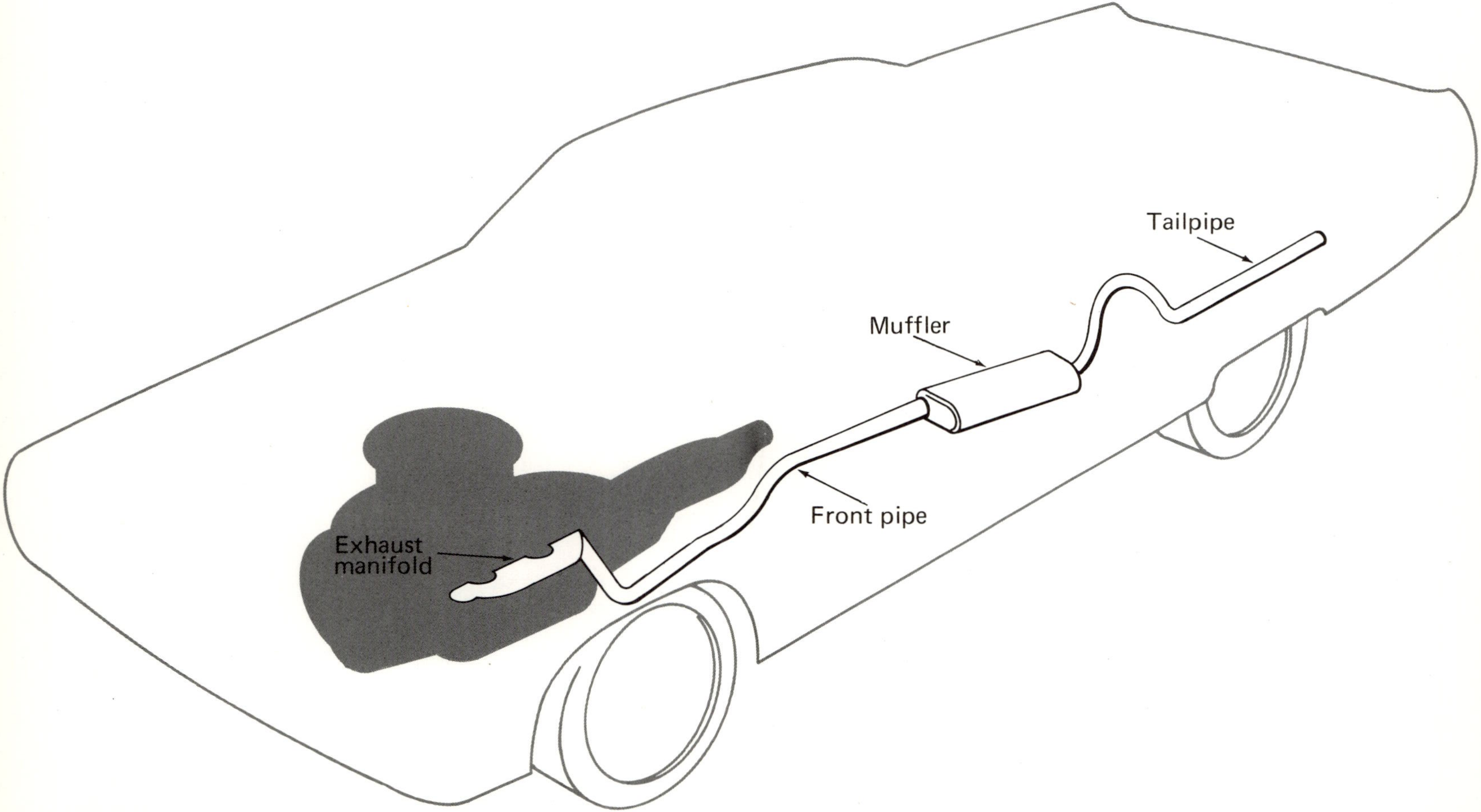
Tailpipe
Muffler
Front pipe
Exhaust
manifold

baffles. As the exhaust gases pass around the baffles, the energy in the hot gases is gradually lost. The sharp pulses of gases that leave the cylinders merge into one continuous flow. There are also *straight through* mufflers. In this type, a perforated pipe runs through the muffler. Glass wool or some other sound-absorbing material is packed around the perforated pipe.

Any muffling system increases back pressure and takes some power from the engine. The design of exhaust systems is really a compromise to bring about as much silencing as possible with the least amount of power loss. The exhaust gases themselves are one of the chief contributors to air pollution. Poisonous carbon monoxide is only one of the polluting gases. In recent years, automobile manufacturers have been directed to install emission-control devices in engines. Among these is the positive crankcase ventilation (PCV) system, which recirculates gases in the crankcase for additional burning.

All of these devices take some power from the engine, reduce fuel economy, and may cause rough idling. However, anyone who has suffered from heavy air pollution in a city will agree to the necessity of these measures.

Widely discussed as possible future pollution-control measures are afterburners in the muffler that would complete the burning of the exhaust gases and emit only carbon dioxide and water. Catalytic converters have also been proposed. These would break down, by chemical action, exhaust products into carbon dioxide and water, and absorb others. The various elements in this system would have to be changed periodically.

# two-stroke and diesel engines

Most engines in early automobiles were two-stroke. They are still widely used in small motorboats, motorcycles, some lawn mowers, and in a few small automobiles, such as some older models of the Saab, built in Sweden.

Two-stroke engines are similar to four-stroke engines in that the essential parts are a piston moving in a cylinder connected to a crankshaft by a connecting rod. There are, however, no valves. Instead, there are intake and exhaust ports that are opened and closed as the piston slides by them. There is one power-producing stroke for every two piston movements. The mixture goes into the crankcase, not directly into the cylinder.

On the power stroke, as the piston goes down, the exhaust port is uncovered and the exhaust gases go out. As the piston goes down, the mixture is also forced from the crankcase into the cylinder through a transfer port, a vertical passage on the side of the crankcase away from the intake and exhaust ports.

When the piston returns on the upward stroke, the inlet port is opened and more mixture enters the crankcase. When the piston

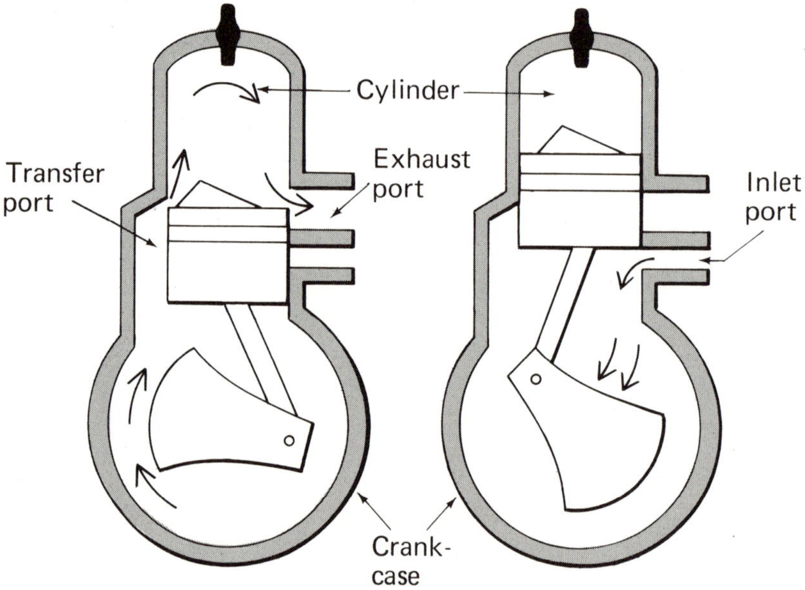


*The two-stroke cycle:*
*left, exhaust and transfer;*
*right, inlet and compression*

reaches the top of the cylinder, the mixture is compressed and ignition occurs.

Theoretically, the two-stroke engine should produce twice as much power per pound of engine as the four-stroke engine. Although, on the basis of weight-to-horsepower comparison, the two-stroke engine is more efficient, it is not *twice* as efficient as the four-stroke engine of comparable displacement. Two-stroke engines can, however, achieve high rpm and are popular with some racing drivers.

The two-stroke engine has the advantage of having fewer moving parts. Since there are no valves, there are no lifters, push rods, and rocker arms, among other parts. But two-stroke engines are lubricated by adding oil directly to the gasoline. This produces a rather dirty, oily exhaust. Two-stroke engines idle roughly with much popping and sputtering.

Large, heavy American automobiles cannot use the two-stroke engine because it cannot deliver the proper power range to drive an automatic transmission. This fact, coupled with the difficulty of designing an emission-control system to handle the oily exhaust, has eliminated the two-stroke engine as a major automobile power plant. But because the two-stroke engine tends to be smaller than the four-stroke, it is useful in such things as power saws and lawn mowers.

The diesel type of internal-combustion engine is named for its chief developer, German engineer Rudolf Diesel. It is similar to the gasoline engine in that it has a piston moving in a cylinder, connecting rods, a crankshaft and a flywheel, and valves operated by a camshaft. There are, however, no spark plugs. Ignition is achieved through the heat generated when air is compressed in the cylinders. Where a spark plug usually is in gasoline engines, there is a fuel injector.

The diesel is a four-stroke engine. On the intake stroke, air is taken in through the intake valve. Air is compressed on the next stroke, and the compression generates heat. On the ignition stroke, fuel is injected into the cylinder, and the heated air causes combustion. Exhaust gases exit through the exhaust valve. Diesel engine pistons usually have a cuplike depression at the top of the cylinder. This helps to initiate combustion by concentrating the fuel.

Diesel engines are used mainly in large trucks and buses, and in locomotives and ships. They are used in construction equipment such as bulldozers and power shovels, and stationary diesel engines are used in industry. There are also a few passenger automobiles equipped with diesel engines.

The high compressions required in diesel engines make them difficult to start. Some are started with small gasoline engines commonly called "pony engines." The exhaust of diesel engines is quite noxious. This is primarily due to the fuels that are burned in them. The fuel is much more "oily" in nature than gasoline and is also much less volatile.

# getting the power to the wheels

In order for an automobile to move, the power that the engine produces has to reach the wheels and turn them. The rotating motion of the crankshaft in most automobiles is transmitted to the rear wheels. Some cars have the motion transmitted to the front wheels, and a few have power transmitted to all four wheels.

The series of mechanisms that transmit the motion to the wheels is collectively called the *power train* or *drive train*. In manual-transmission (gear-shifting) automobiles, it consists of the *clutch*, the *gearbox*, the *drive shaft*, and the *final drive*, or *rear-end*. At the final drive, the horizontal rotation of the crankshaft is changed to the rotary motion of the wheels. The power train is designed so that the nature of the power delivered to the wheels can be changed to meet different driving conditions.

• *The clutch.* The clutch provides a means of disconnecting and reconnecting the engine to the power train. It would obviously be inconvenient to turn the engine off every time the driver wanted to stop the vehicle. And the disadvantage of starting the engine at

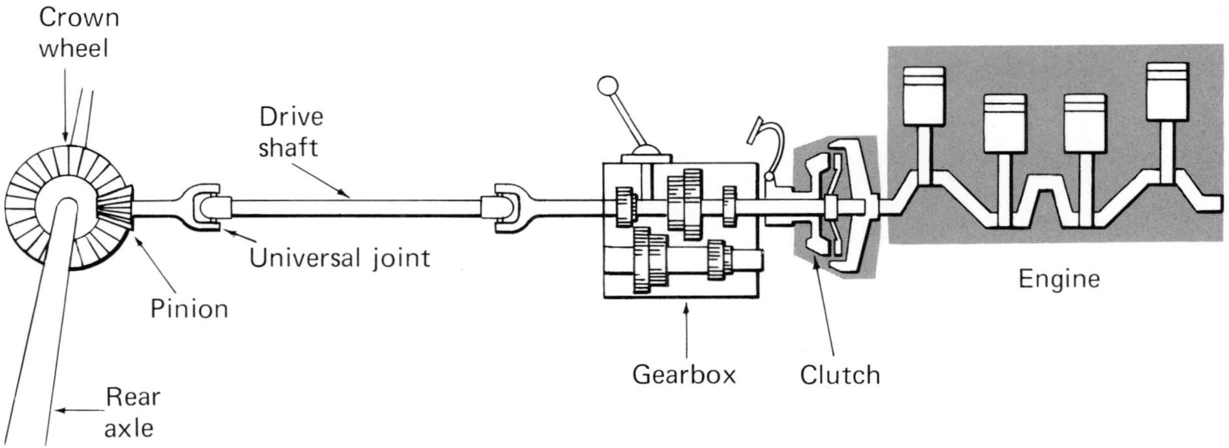

Crown wheel
Drive shaft
Universal joint
Pinion
Rear axle
Gearbox
Clutch
Engine

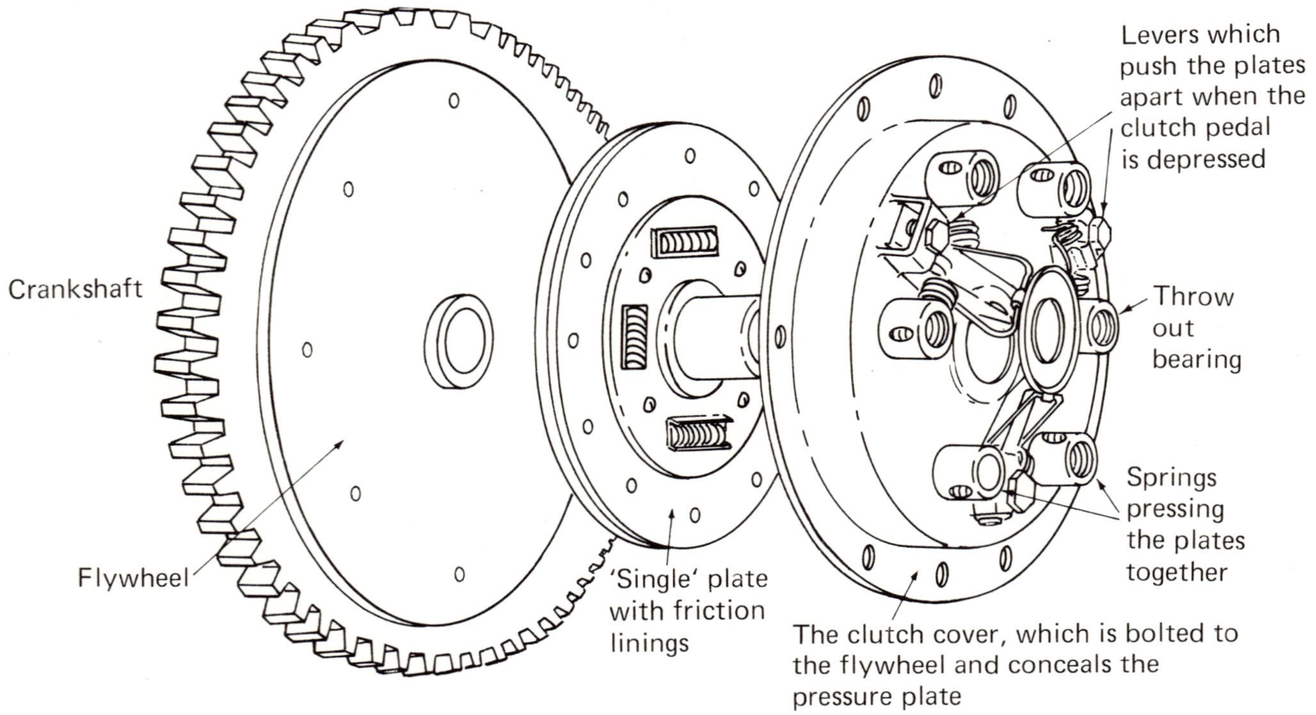
Levers which push the plates apart when the clutch pedal is depressed
Crankshaft
Throw out bearing
Springs pressing the plates together
Flywheel
'Single' plate with friction linings
The clutch cover, which is bolted to the flywheel and conceals the pressure plate

the same time the engine is delivering power to the wheels is also obvious.

Most manual-transmission cars are equipped with what is called the *single dry plate friction clutch.* The name is somewhat misleading since there is more than one "plate" in the clutch assembly. There are essentially three plates: flywheel, friction plate, and pressure plate. A clutch plate cover is over the assembly. When these plates are together, there is friction. One plate turns the other; motion is transmitted to the gears and, ultimately, to the wheels. When there is a gap between the plates, no motion is transmitted.

The flywheel plate is securely bolted to the flywheel. As long as the engine is running, this plate is in motion. The engagement of the clutch plates must be gradual. If the plates come together abruptly, the car might bolt forward like a projectile. This would be uncomfortable and even dangerous for the passengers. Smooth starting is accomplished by pressure gradually pressing the friction plate against the flywheel.

The clutch is controlled by a pedal in the driver's compartment. When the pedal is pushed down, spring levers push on a *throw-out bearing,* which pulls the plates apart. When the pedal is allowed to rise, heavy springs between the cover and the pressure plate push the pressure plate against the friction plate and flywheel. A shaft leading from a tubular connection on the friction plate passes through a hole in the clutch assembly cover to the gearbox. At first contact, there is some slipping, but as the pressure is increased, the contact becomes firmer and all plates rotate at the same speed.

Friction plates are usually made of an asbestos material. They wear with use and eventually must be replaced. The friction produces heat, which is usually removed to the outside air. If the pressure holding the plates together is not strong enough, the resultant slipping can cause enough heat to burn the friction plates.

The clutch is usually disengaged when the car is stopped and when the gears are shifted. If the clutch is not disengaged on stopping, the engine will usually stall.

• *The Gearbox or Transmission.* Motion is transmitted from the clutch to the drive shaft by a set of gears. The arrangement of the gears can be altered to achieve different speeds and *torques* to meet different driving conditions. Torque is the term for the force applied by a turning object, such as a gear. Gears are metal wheels or disks that have evenly spaced teeth around the circumferences. The teeth enable one gear to mesh with another so that one gear can turn another one. In general, a slower moving gear has more torque than a faster moving one. When the speed of a turning object is increased, there is less torque. More torque is required to overcome the inertia of a stationary automobile and get it moving than is required to keep it moving. In other words, more torque is needed to increase the speed of the vehicle from 0 to 20 miles per hour than from 60 to 80 miles per hour. There must, however, be a means of readily changing the gear arrangement.

• *Gear ratios.* The gear that receives the engine power is the *driver* gear. The gear it turns is the *driven gear.* If the driver gear is larger than the driven gear, the speed of the driven gear will be greater. Its torque will be less. If the driver gear is smaller than the driven gear, the speed of the driven gear will be less, but its torque will be greater. Assuming that the gear teeth are equal in size, the smaller gear will, of course, have fewer teeth.

The comparison of the number of teeth in the meshing gears is the *gear ratio.* If the driver gear has 20 teeth and the driven gear has 60 teeth, the ratio is 3:1. This is *reduction gearing,* since the speed is reduced. When the speed is increased, *multiplication gearing* takes place. In an automobile gearbox, there are of course more than two gears. The gear ratio is the product of the number of teeth in the driven gears compared to the product of the number of teeth in the driver gears. For example:

$$\frac{\text{driven gears: } 30 \times 15}{\text{driver gears: } 15 \times 20} = 1.5 \text{ to } 1 \text{ (reduction)}$$

• *Gear Shifting and Manual Transmission.* Most cars are equipped with a *sliding gearbox.* The gears are mounted on shafts. There are two or more shafts of gears. The number of gear ratios (commonly referred to simply as "gears"), or speeds, is generally three or four forward and one reverse. Some racing and sports cars have five forward, and some heavy trucks have as many as twenty. In three- and four-speed gearboxes, there are usually two shafts of gears. The shaft that leads from the clutch (the main shaft) is split into two shafts, the first motion and output. These are the sliding shafts. The *layshaft* does not slide and is usually below the sliding shafts. These gears are also called *constant mesh*, since they are always in mesh. (See the diagram on page 62.)

The gearshift lever is connected to a *selector shaft*. The selector forks extend from the selector shaft. When the gearshift lever is moved, the gears on the main shaft slide back or forward to mesh with gears on the layshaft. There is usually one gear on the first-motion shaft, and it is in constant mesh with the first gear on the layshaft. This is a reduction gearing and provides added torque.

The gear ratios obtained vary with the requirements of the vehicle. Generally, there is a first, or "low," gear ratio, which provides maximum reduction and torque, one or two intermediate reduction gears, and a high gear, which is direct — that is, no reduction or multiplication. Some gearboxes have overdrive, which is a multiplication gear. This is used only at high speeds and allows the gears and driveshaft to turn faster than the crankshaft. It reduces engine wear, but requires special modifications in the final drive.

Following are the paths of the transmitted rotation in the various gear ranges in a three-speed transmission. (See diagrams on page 63.)

*Neutral:* The main shaft is separated from the first-motion shaft, and motion is transmitted only to the layshaft. No motion is transmitted to the wheels. "A" and "B" gears are always in mesh.

*First, or Low, Gear:* Gear "G" moves forward to mesh with gear "E" on the layshaft. A typical ratio is 3.5:1.

*Intermediate (Second) Gear:* Gear "B" on the main shaft slides back to mesh with gear "E" on the layshaft.

*High Gear:* Gear "B" slides forward along the main shaft to join the first-motion shaft. "Dogs" extend forward from gear "B" to mesh with dogs at the end of first-motion shaft. The ratio is 1:1.

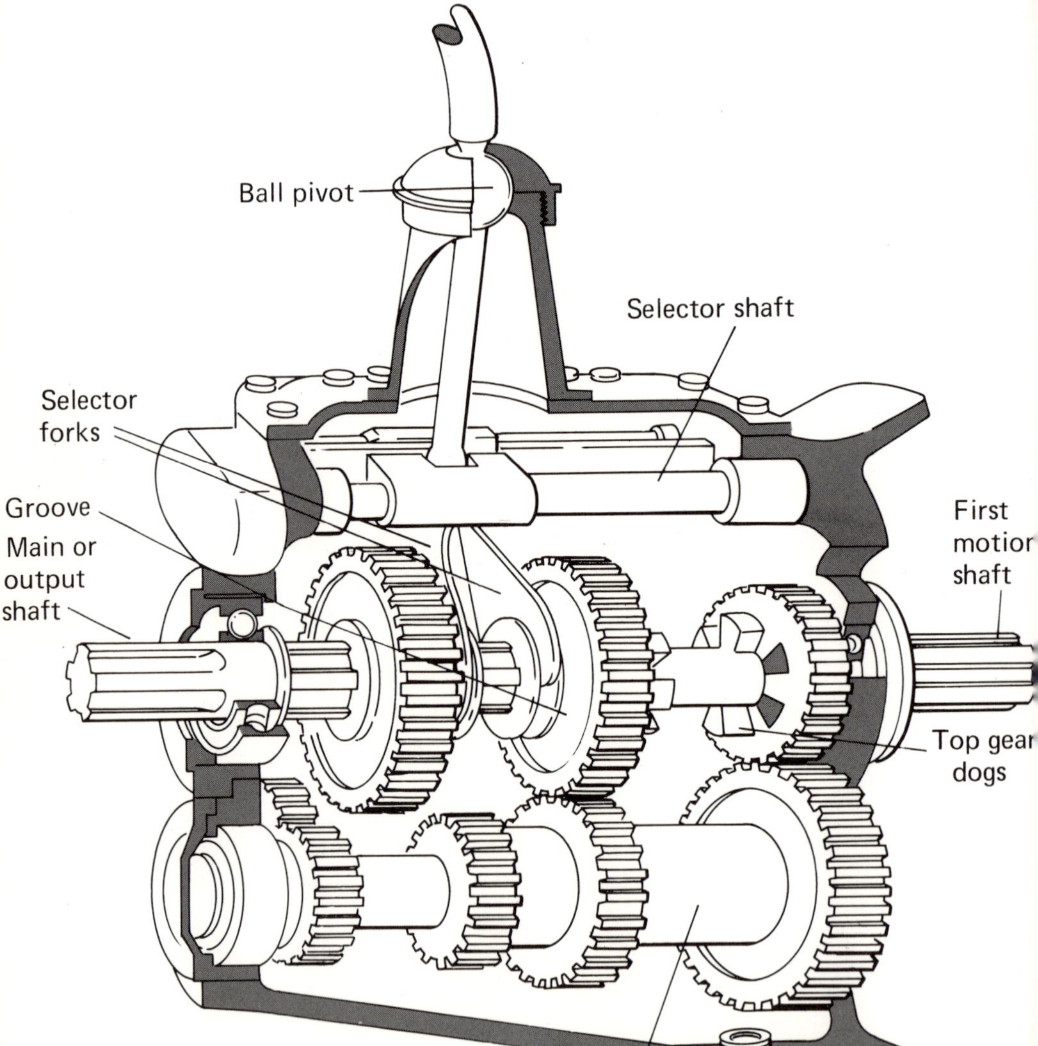


*Gear ranges in a
three-speed transmission*

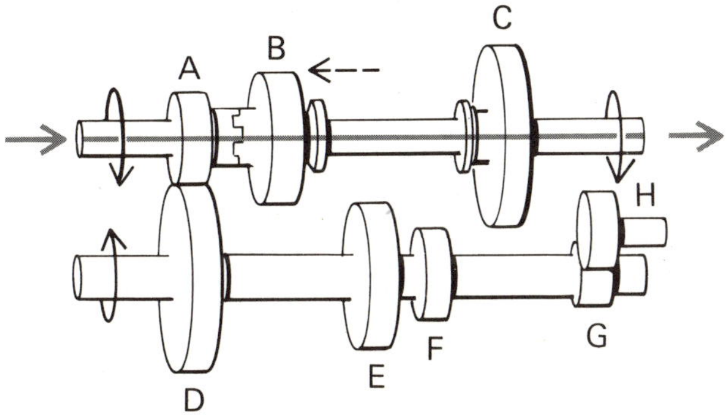


High gear

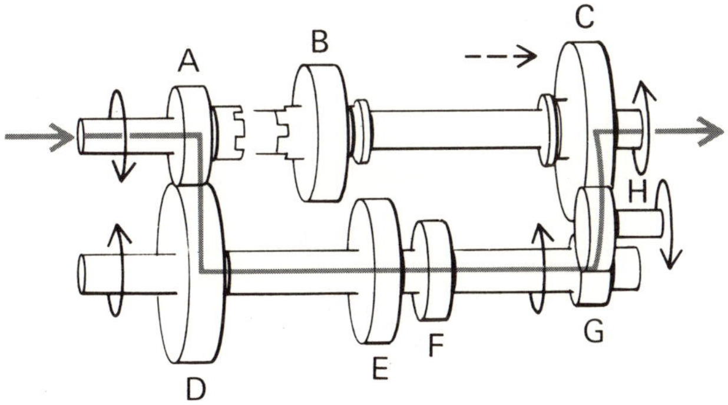


Reverse gear

*Reverse Gear:* Gear "G" is always in mesh with "H," which is the *reverse idler gear.* Gear "H" rotates on a short shaft fixed on the gearbox casing. To shift into reverse, gear "G" slides to mesh with "H." This changes the direction of the main shaft. Reverse gear ratios are generally around 4:1.

• *Synchromesh.* The purpose of changing gears is to enable the proper torque to be delivered to prevent straining the engine when starting from a dead stop, accelerating, and climbing hills. This requires that gears be shifted while the car is moving. In order to do this without clashing and grinding of gears, the gears that are being brought into mesh must be turning at the same speed. Although a very skilled driver might be able to shift silently with nonsynchromesh gears, most drivers cannot, so practically all cars with manual gearshift have synchromesh. It allows silent, synchronized gear shifting in the moving gears. Some units allow synchronized shifting in all forward gears, including low. Usually no synchromesh is provided for reverse, since this gear is never to be engaged while the car is moving. Gears in synchromesh gearboxes are usually helical — cut at an angle. This type of gear runs more quietly than straight-cut gears.

The gearbox illustrated is a three-speed box with second and high gears synchronized. The synchromesh unit also has a layshaft and first-motion and main shafts. The engaging point of the first-motion and main shafts has an outer steel hub around the engaging dogs. The engaging dogs are somewhat like keys that fit together. The selector fork is fitted to the outside hub. An inner hub that has internal and external engaging dogs is free to slide on the main shaft. The whole unit rotates with the main shaft. The inner hub has internal cones at each end.

The hubs are held together with ball and springs located in the inner hub. The balls are pressed into a groove around the center of the bore of the outer hub. When the gearshaft lever is moved, the balls are pressed against the springs so that the outer hub can slide over the inner hub. On the inner face of gears "A" and "C" are rows of engaging dogs.

Refer to the illustrations to follow an example of synchronized shifting. In shifting from first to second, there is, as in all gearshifts, a passage through neutral. As the gearshift lever is moved, the synchromesh unit slides along the main shaft and brings the cone of the inner hub into contact with the external cones on gear "B." The cones are pressed together, and the speed of gear "B" is synchronized with the speed of the main shaft.

Continued movement of the gearshift lever relieves the pressure of the springs on the balls and pushes the internal "keys" of the outer hub into mesh with the dogteeth on gear "B." This gear is now locked to the synchromesh unit. Motion will pass through gears "A" and "D," the layshaft, gear "C" to gear "B," and through the hubs of the synchromesh unit to the main shaft.

• *Automatic Transmissions.* There are many types of automatic transmissions. The variations in design are a matter of cost and the type of vehicle in which the transmission is used. A more

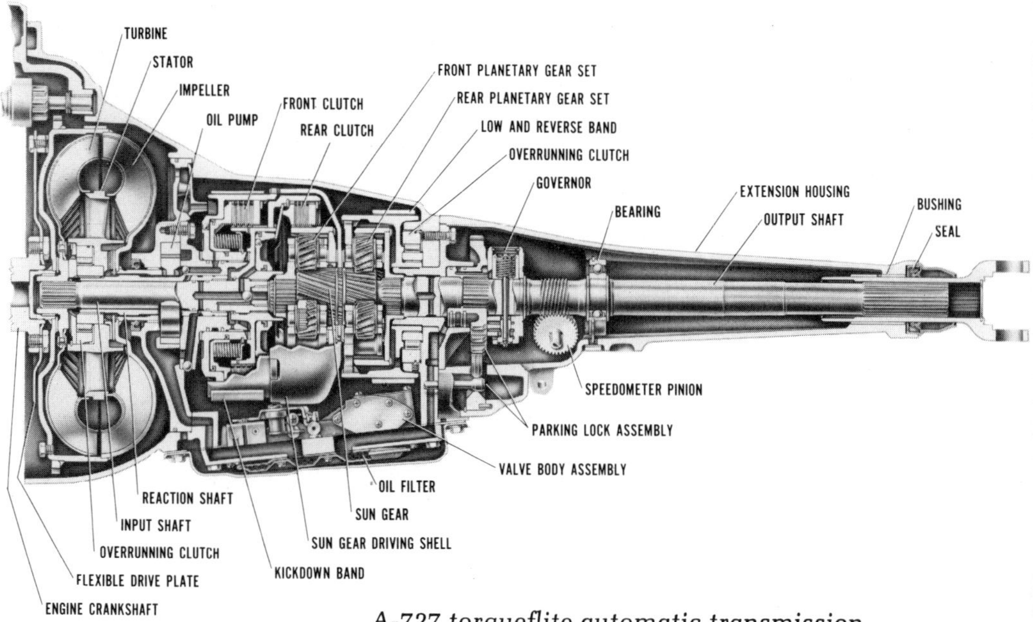


*A-727 torqueflite automatic transmission*

or less typical unit will be described here. The major components are some type of *automatic clutch* in conjunction with a *torque converter*, and a set of gears. Automatic transmissions are made in two, three, or four speeds and with several ranges for cruising, hill climbing, and other extreme conditions, such as snow or muddy roads.

The several types of automatic clutches include electric, pneumatic, and centrifugal. The usual type, however, is the *fluid clutch,* also called the *fluid flywheel* and the *fluid coupling.* Basically, it is a set of opposed vanes, somewhat like windmills. The vanes are enclosed in a housing, and the housing is filled with oil. One set of vanes is turned by the engine. As it turns it sets the oil in motion, and the moving oil gets the other set of vanes moving. The vanes attached to the crankshaft are the *driving member;* those that ultimately transmit power to the wheels are the *driven number.*

The oil transmits power smoothly and continuously. It is not necessary to disengage the fluid clutch on stopping. When the engine is running at idle, very little force is transmitted from the driving to the driven member. The force of the brakes on the wheels is enough to hold the car stationary. As the speed of the engine is increased, the force transmitted to the driven member is increased. The fluid coupling cannot increase torque. For this, a torque converter and gears are required.

The torque converter is built into the fluid-clutch mechanism. It serves somewhat the same function as a gearbox. However, it increases torque continuously rather than in specific steps, as is the case in a manual gearbox.

In the torque converter, the *stator* is added to the driving and driven member. The driving and driven members are called the *impeller* and *turbine* respectively. The impeller drives oil against the turbine as in the simple fluid clutch. If the speed of the turbine is much less than that of the impeller, the oil will rotate backward after leaving the turbine, in relation to the casing. The stator blades change the direction of this backward rotation. The torque on the turbine may be double that on the impeller. Maximum torque is obtained when the engine is not moving. As the engine speed and vehicle speed increase, the torque decreases. When the impeller speed is a little greater than the turbine speed, the torque converter acts as a fluid flywheel, and torque increase ceases.

Up to a point, in lighter vehicles, a torque converter could be used to transmit power to the wheels without the use of a gearbox. Different converters could be mounted in tandem. Each converter

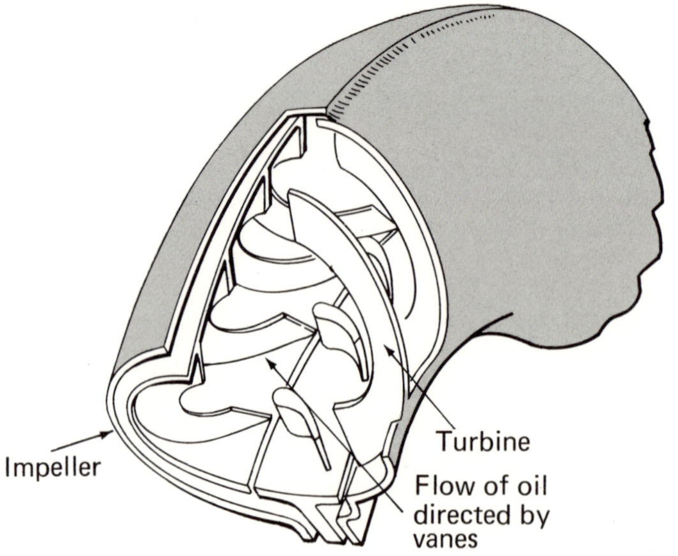

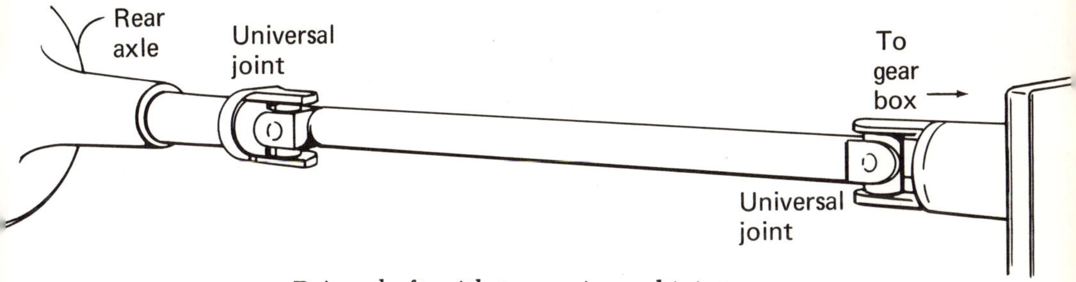


*Drive shaft with two universal joints*

would be constructed to provide different degrees of torque to correspond with low and high gears. This arrangement, however, would be very bulky, and torque converters are usually linked to a gearbox.

The gearbox can be of the sliding type, but most automatic transmission gearboxes have *planetary* gears. In this system, there are larger "sun" gears, with "planet" gears meshing around them. The automatic transmission gearbox is equipped with speed detector devices that automatically change gear ratios as the speed of the car increases or decreases.

The range of the gear ratios is changed by a lever, usually mounted on the steering column. The lever in most cars has an indicator with letters such as P R N D L. P is "park," which is a brake and is engaged only when the car is not moving. R is "reverse," N is "neutral," D is "drive," for normal cruising, and L is "low." Some may have two low ranges and two drive ranges — one for city driving and one for the highway. Since the mid-1960s, the order of the ranges on the indicator has been standardized in all automobiles. For reasons of safety, N is always between R and D, to lessen the chance that a driver might select R when he thought he was selecting D and vice versa. When the selector level is moved, the gears are rearranged to effect the desired range.

• *Drive Shaft.* Most cars have the engine in front and the driving wheels in the back. A rotating shaft — the drive shaft — transmits the motion from the gearbox main shaft to the final drive in the rear axle.

The drive shaft is a tube of metal fitted with one or more *universal joints* and sometimes a *sliding joint.* The universal and sliding joints give flexibility to the drive shaft so that it can be

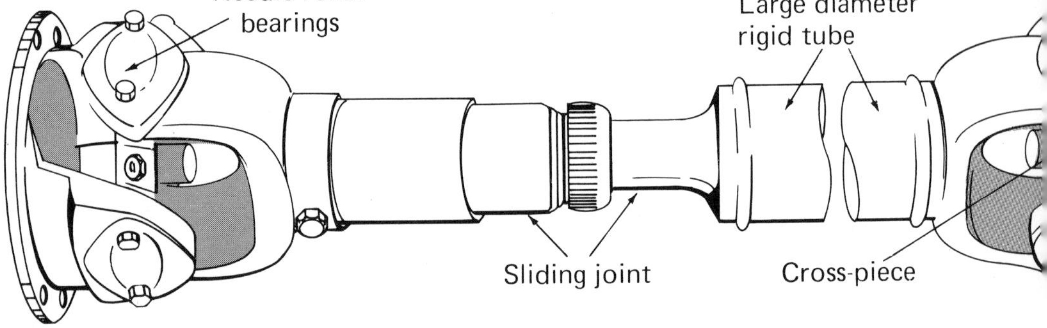


*Drive shaft with two universal joints and a sliding joint*

accommodated to road shock. The rear wheels move up and down over bumps and ruts. If the drive shaft could not "give" in response to these shocks it might break. There would also be an uneven delivery of torque to the rear end.

Universal joints are usually mounted on the shaft at the ends near the main shaft and the rear-end assembly. Some cars have another one at some point between the ends of the shaft. There are many different designs for universal joints, but in all of them, the idea is to effect uninterrupted rotary motion while allowing the shaft to move in any direction (up, down, sideways, or angled). In a simple type of construction, each end of the shaft is forked and pivots on a common crosspiece. Others employ "needle rollers" to reduce friction. Sliding joints, as the name implies, allow slight telescoping action in the shaft.

• *Final Drive — Rear-Axle Assembly.* At the rear end of the drive shaft is a cone-shaped (actually a truncated cone) gear called the *pinion gear.* It meshes with (and transmits torque to) a vertically mounted circular gear called the *ring gear.* The rear axles project from the ring gear. The mesh of the pinion and ring is at right angles, so that the motion of the ring gear is in the same plane and direction as the rear wheels. In cars with rear-wheel drive, the vehicle is pushed forward by the motion of the wheels. The final driving force is actually the friction of the wheels against the road surface. The rear-axle assembly also acts as a reduction gear between the drive shaft and the road wheels.

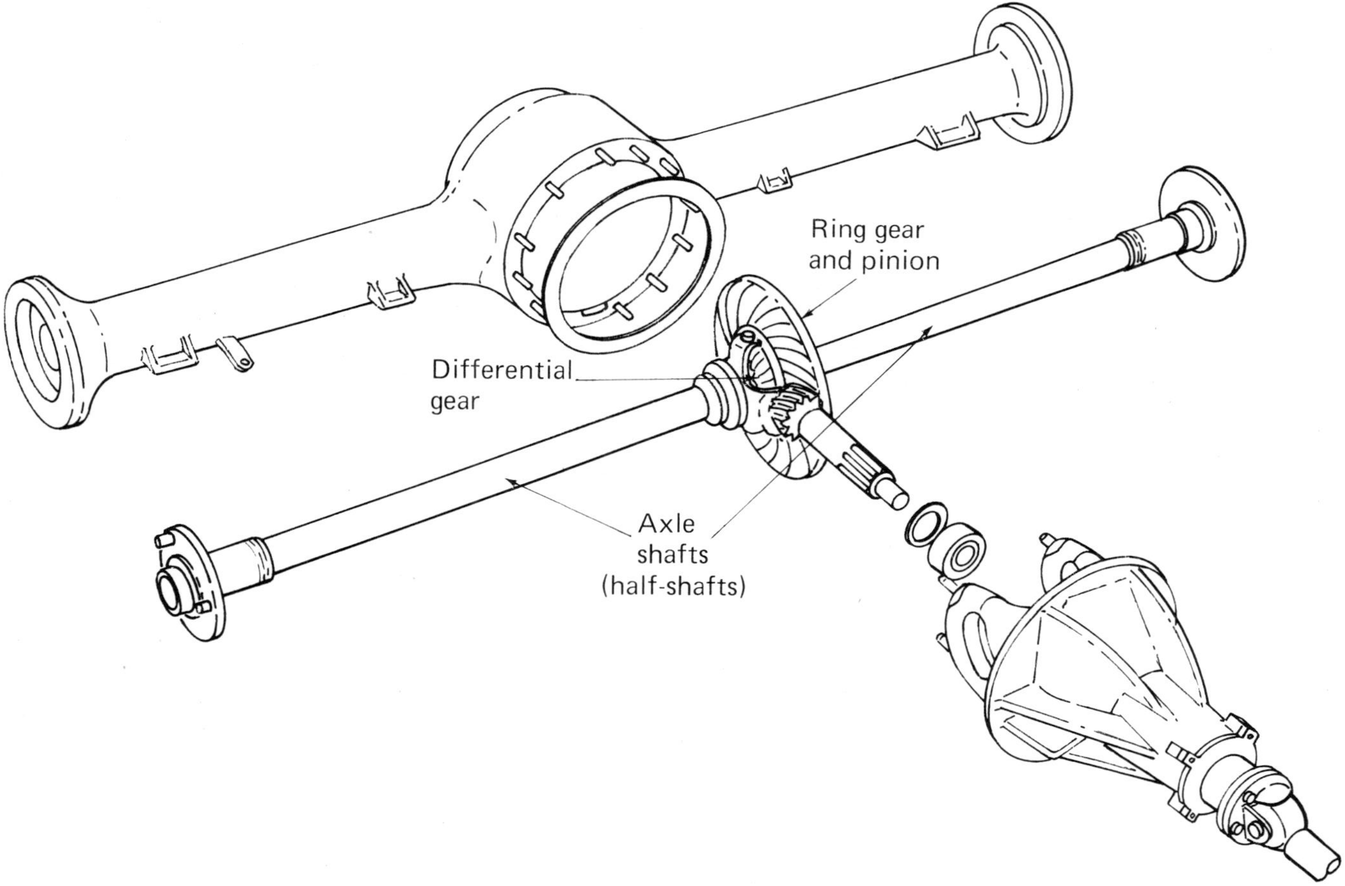
Ring gear
and pinion
Differential
gear
Axle
shafts
(half-shafts)

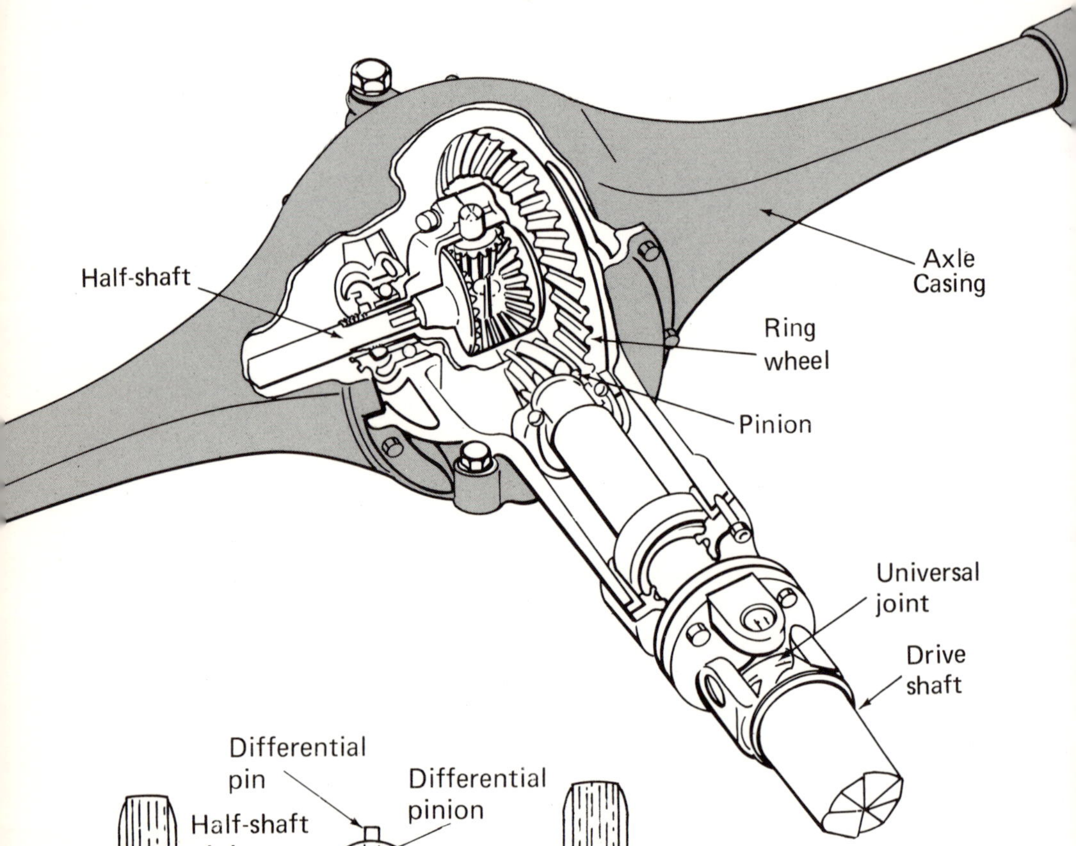
Half-shaft
Axle Casing
Ring wheel
Pinion
Universal joint
Drive shaft

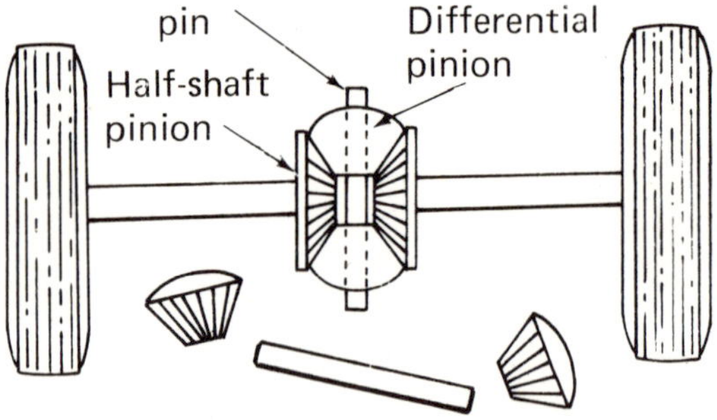
Differential pin
Differential pinion
Half-shaft pinion

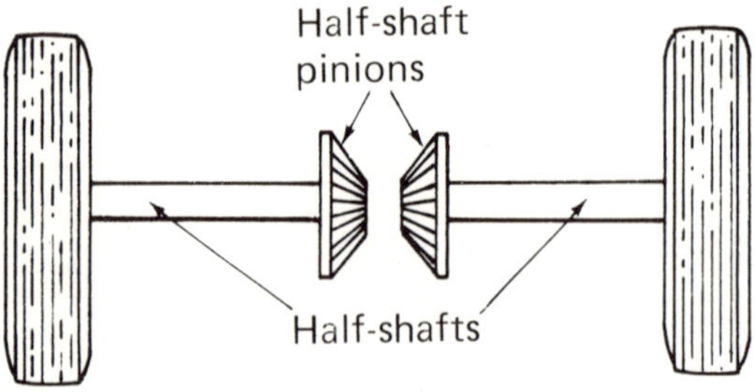
Half-shaft pinions
Half-shafts

The gearing on the ring and pinion is usually one of two different types — *spiral bevel* and *hypoid*. In spiral-bevel gearing, the pinion meshes with the ring at the center line; that is, the widest diameter plane of the circle as it is mounted in the rear-end assembly. In hypoid gearing, the mesh is below center line. This allows the drive shaft to be lower, and therefore allows more room for passengers by reducing the height of the "hump," or drive shaft, tunnel in the passenger compartment.

The ratio between the ring and pinion is, in most vehicles, fixed and is chosen by the manufacturer as the one best for the general use of the car. Some heavy trucks have two rear-axle assemblies geared at different ratios. One or the other can be used as the driving situation demands.

• *Differential.* Included in the rear axle is a differential, which enables the wheels to rotate at different speeds when the car is turned around a corner. When a car turns a corner, the wheel on the inside of the turn (the left wheel on a left turn) travels a shorter distance than the wheel on the outside, which has to travel a greater distance. The torque delivered to each wheel, however, must be the same. This seemingly contradictory set of mechanical events is accomplished by the differential, which is mounted in the "hole" in the ring gear. The ring is bolted to the differential housing so that the differential turns with it. The parts of the axle that extend from the rear-end assembly to the wheels are called *half-shafts (half-axles)*. The sequence of mechanical connections from the half-shafts is: each half-shaft meshes with the *differential pinion gears*, which are free to rotate on a *differential pinion* (pin or shaft). All four pinion gears are enclosed in a differential housing. The entire assembly, including the half-shafts, rotates as one unit. When the vehicle is moving in a straight path, the two half-shafts are locked and they rotate as a single unit.

When the vehicle is turned, the inner wheel slows down and the outer wheel moves faster. The differential pinion will rotate on the pin with the pinion rolling around the pinion gears mounted at the inner ends of each axle. The ring gear will rotate at a speed that is the average of the speed of the two rear wheels. The torque transmitted to each wheel remains the same because the rotating pin is at the same distance from each half-axle and transmits the same driving force to each half-wheel pinion gear.

Differentials can be troublesome on icy pavements. If one wheel is spinning on ice, it will be turning rapidly (but going nowhere) and consequently the other wheel will receive hardly any driving force. "Limited-slip" (sometimes called "nonslip") differentials, which are offered under a variety of trade names by different manufacturers, overcome this problem. The mechanical details differ, but basically there is a set of clutches that connect the differential housing to the half-axle pinions. These clutches are activated when the icy spinning conditions are encountered.

• *Front-Wheel Drive.* Supplying the drive to the front wheels instead of the rear has many advantages, but it also presents mechanical problems. With front-wheel drive, the car is *pulled* around corners rather than *pushed,* and this increases stability. Road holding is improved, and a front-wheel-drive car is less likely to skid. Since there is no drive shaft, there is no need for a "tunnel," and so there is more room for the passengers. Such cars can be lower, and the lower center of gravity increases stability. Traction in snow and ice is improved by the weight of the engine on the front wheels.

The mechanical problems arise from having the drive in the steering wheels. When the wheels are turned and/or go over bumps, there is a tendency for the power to be transmitted to the wheels in spurts rather than in a smooth flow. In rear-wheel-drive cars, the universal joints compensate for the road shocks. A special type of universal joint is one of the mechanical devices that overcomes the problems of front-wheel drive. This is the *constant velocity universal joint.*

The constant velocity universal joint delivers torque at even rates at whatever angles the wheels are turned. There is, of course, a differential to allow for turning corners. The joints are located at the ends of the axles just at the wheels. A series of steel balls arranged in a circle around the end of the axle receive power. The driven part attached to the wheel covers the ball assembly. A spherical ball cage locates and controls the position of the balls as the wheels are turned, so that the centers of the balls are always in a plane that bisects the angle between the driving and driven parts of the axle. This equalization of angles keeps the velocity of the wheels constant in reference to the action of the differential.

• *Four-Wheel Drive.* Drive transmitted to all four wheels is useful for vehicles designed to run cross country, and in deserts, hilly places, and other areas with bad roads or even no roads. There is usually a control lever in the car with which the driver can connect or disconnect the drive to the front wheels. The control is linked to an additional gearbox, from which drive shafts extend to both sets of wheels. In most four-wheel-drive vehicles, the gear ratios supplying torque to either set of wheels can be changed.

# driving and stopping—brakes

As the speed of automobiles increased over the years, the efficiency of the brakes had to be improved. Brakes in early automobiles were much like the brakes on horse-drawn wagons — a piece of leather or rubber that was brought to bear on one or more wheels. Brakes still work on the friction principle, but the friction force is applied to the inner surface of the wheel rather than the outer surface.

Brakes convert the energy of motion of the moving automobile to heat energy. The heat energy is lost to the air, and the car slows down. If the braking action is continued, the car stops. In hard breaking, heat sometimes builds up at a faster rate than it is removed. When this happens, there is a loss of efficiency and the brakes are said to fade. Fading is diminished by designing brakes to present as much surface for heat loss as possible.

Since the mid-1930s, all automobiles have had *hydraulic* brake systems. The force from the brake pedal is transmitted to the compression of fluid in a closed system of cylinders and lines. In earlier automobiles, the force was transmitted mechanically by wire cables.

• *Drum Brakes.* These are the most frequently used system, although disc brakes are gaining in popularity. A *brake drum,* which is a hollow metal cylinder, is fitted to each wheel. It is an integral part of the wheel and rotates with it at the same speed. Fitted inside each wheel drum are *brake shoes.* These do not rotate and are not in contact with the drum when the car is moving. Each brake shoe has a *lining,* usually made of some asbestos material. There are usually two brake shoes in each wheel.

The shoes are attached to a hydraulic *wheel cylinder,* a return spring, and an adjuster. When the driver steps on the brake pedal, the pressure of the fluid in the wheel cylinder forces the linings against the inner surface of the brake drum. It is rather important that the braking force be applied evenly to all wheels. Uneven braking can result in swerving, unexpected twists of the front wheels, and, in extreme cases, complete loss of control. The hydraulic system is a simple but effective system for even distribution of braking force.

In hydraulic systems, there is a *master cylinder,* filled with fluid. Tubes or lines extend from the master cylinders to the wheel cylinders. The entire system — master cylinder, lines, wheel cylinders — is completely filled with fluid, somewhat like a light oil.

The brake pedal is linked to a piston in the master cylinder. The piston applies pressure to the fluid. When the fluid is pressed, the pressure is transmitted to each of the wheel cylinders, and this pressure pushes the linings against the drum. When the brake pedal is released, a spring pulls the linings back.

Brake linings wear, and as they wear they must be adjusted. Eventually, they must be replaced. Each wheel has an adjuster that can be reached through an access hole in the drum. Some cars have self-adjusting brakes. This mechanism allows the brake linings to move a little closer to the drum as wear takes place. In many systems, the self-adjustment takes place when the car moves in reverse.

Most automobiles have a dual braking system, so designed that if one system fails, the other will supply some braking until repairs can be made. If all the fluid leaks out, there can be complete loss of braking. There is a reservoir of brake fluid in the master cylinder that adds fluid to the system if any is lost. The level of this reservoir should be checked from time to time. Most drivers, however, seldom remember to do this. But if there is a leak, the loss is usually gradual enough so the driver will notice

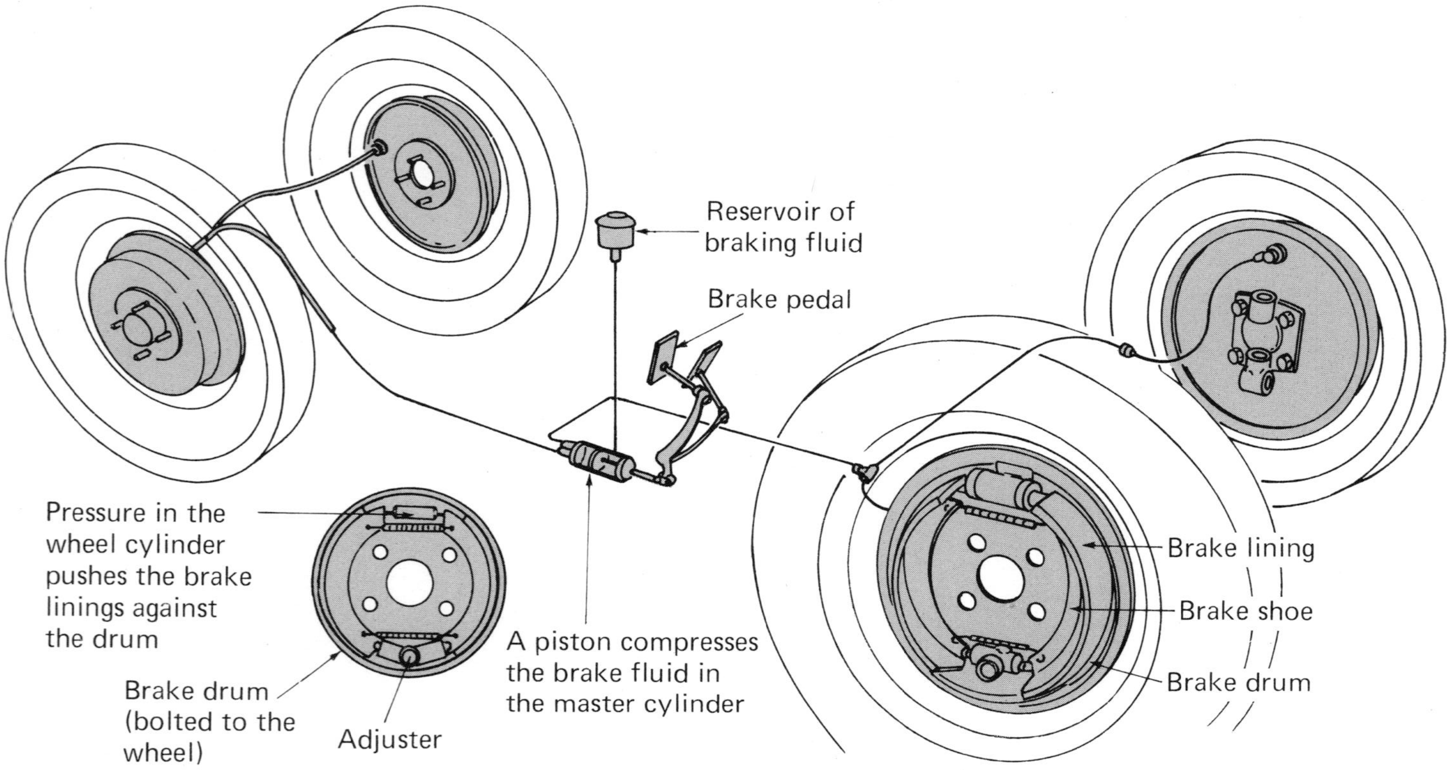
Reservoir of braking fluid
Brake pedal
Brake lining
Brake shoe
Brake drum
Pressure in the wheel cylinder pushes the brake linings against the drum
A piston compresses the brake fluid in the master cylinder
Brake drum (bolted to the wheel)
Adjuster

that his brakes are not quite right, and he might then have them serviced. Air in the system makes the brakes feel spongy and unresponsive. *Bleeder valves* are on each wheel cylinder. Fluid can be bled off at these valves to remove air bubbles.

• *Disc Brakes.* In this system, there are no drums or shoes. A metal disc rotates with the wheel. The control is also hydraulic. Brake shoes are replaced by a set of lined *calipers.* When braking pressure is applied, the lined calipers clamp against the disc. The action is much the same as bicycle hand brakes, which are blocks that rub against both sides of the wheels.

*Floating-caliper disc brake*

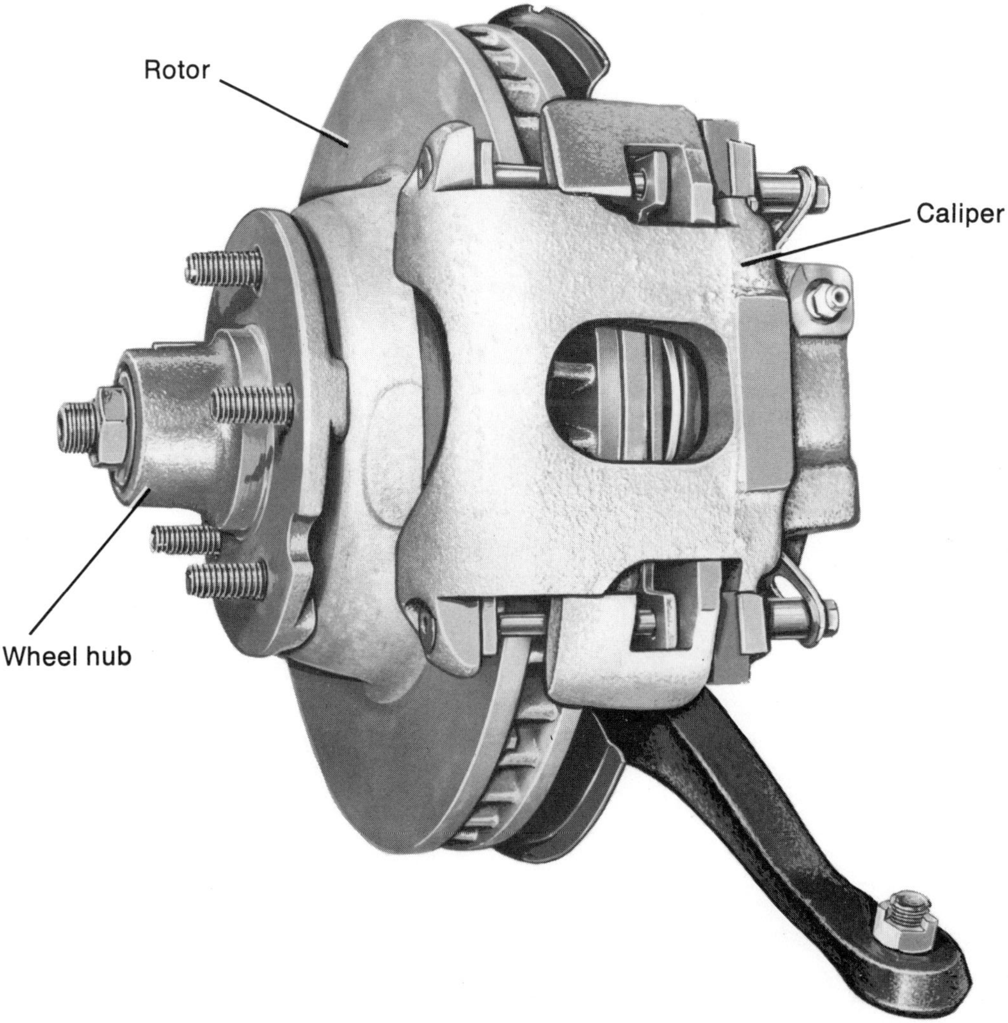

Disc brakes tend to be more effective than drum brakes. Since the caliper is in contact with only one part of the disc during braking, the rest of the disc is free to give off heat. This reduces fade. The caliper linings in general wear more evenly than shoe linings, so that over the period of use, braking force tends to be more even and consistent. Disc brakes, however, are more expensive to buy and service, but as their use increases, this may change. Many automobiles are available with disc brakes on all four wheels or with disc brakes in the front wheels and drums in the back. There is much less likelihood of wheel "pull" (twisting to the right or left) with disc brakes in the front wheels.

• *Power Brakes.* As automobiles were made larger, heavier, and faster, greater foot pressures on the part of the driver were required for braking. This was a potential safety hazard. Since the mid-1960s, almost all larger, heavier American automobiles have been fitted with power brakes. Power assist can be used with both drum and disc brakes.

There are various power brake mechanisms. Almost all utilize a difference in pressure between the atmosphere and the partial vacuum in the intake manifold. The difference in pressure gives an additional boost to the master cylinder piston. Brakes are generally applied when the foot is off the accelerator pedal. Very little fuel enters the cylinders at that point, and the vacuum in the intake manifold is fairly high.

A fairly large piston is installed in a cylinder called the *slave cylinder.* A vacuum is on one side of the cylinder and air on the other. A plunger on the piston pushes into fluid in the master cylinder or a passageway leading from the master cylinder.

When the brake pedal is pressed, a valve opens to allow air into the cylinder on the side of the piston away from the piston plunger. Since there is a vacuum on the other side, the piston moves with great force to apply pressure on the brake fluid system. A heavy spring pushes the piston back when the brake is released. The only effort required of the driver is enough pressure to open the valves.

When the brake pedal is first applied, the valve that admits air to the system is closed, the vacuum valve is opened, and there is a vacuum on both sides of the piston. As the pedal is pressed farther, brake fluid opens the air valve to let air in until the air

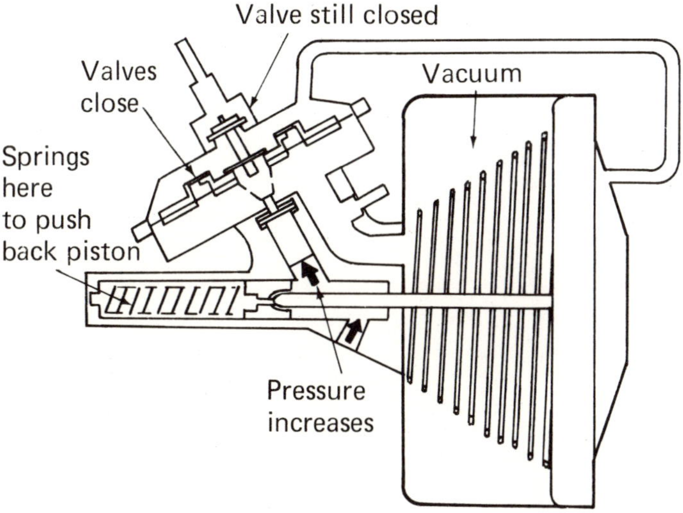
Valve still closed
Valves close
Springs here to push back piston
Vacuum
Pressure increases

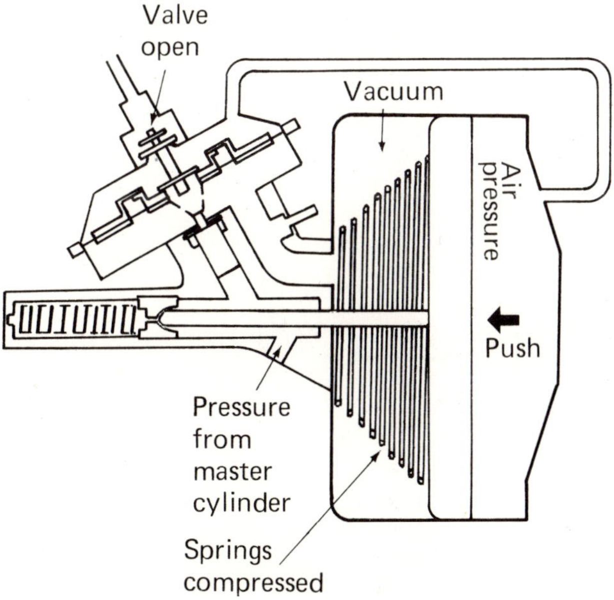
Valve open
Vacuum
Air pressure
Push
Pressure from master cylinder
Springs compressed

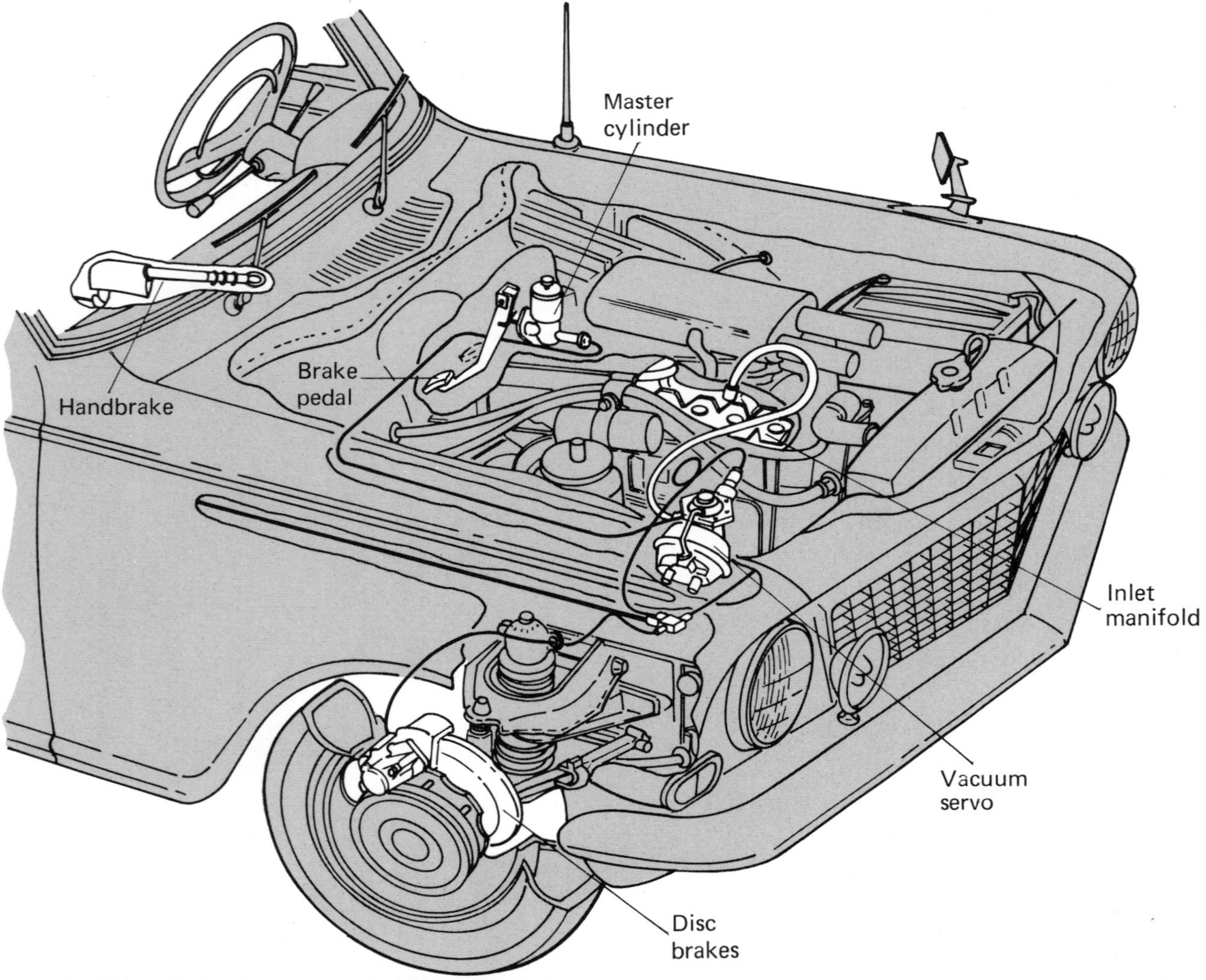
Master
cylinder
Brake
pedal
Handbrake
Inlet
manifold
Vacuum
servo
Disc
brakes

pressure is equal to the pressure of the fluid acting on the valve. Air is then available to push the piston. The spring return of the piston restores the vacuum condition.

If the power system breaks down, the brakes still work, but, of course, more foot pressure will be required. The energy to activate power brakes and steering comes from the engine, so the power boost is not available if the engine is not running.

*In this automobile, a vacuum servo operates disc brakes on the front wheels and drum brakes on the rear wheels. In this kind of power braking, a vacuum is created on one side of a piston and is pushed from the other side by air pressure.*

# steering

The design of efficient steering was a particularly difficult problem for early automobile makers. The efforts of engine designers to make faster, more efficient engines outran the development of safe and efficient steering systems in the early years. The earliest cars had steering tillers hooked directly to a swing bar on which the front wheels were mounted. It worked at low speeds, but a good road bump could wrench the tiller from the driver's hand. If the driver got away with a bruise he was lucky. Broken wrists, arms, ribs, and even jaws were frequent results.

The main idea in steering is to link the driver's effort on the steering wheel to the front wheels. Large movements on the steering wheel are translated to smaller movements in the wheels. Various gear arrangements are used to achieve this. In turning, the inner and outer wheels travel on circles of different radius, so the steering mechanism must deliver a sharper angle of turn

*Lever arm steering system*

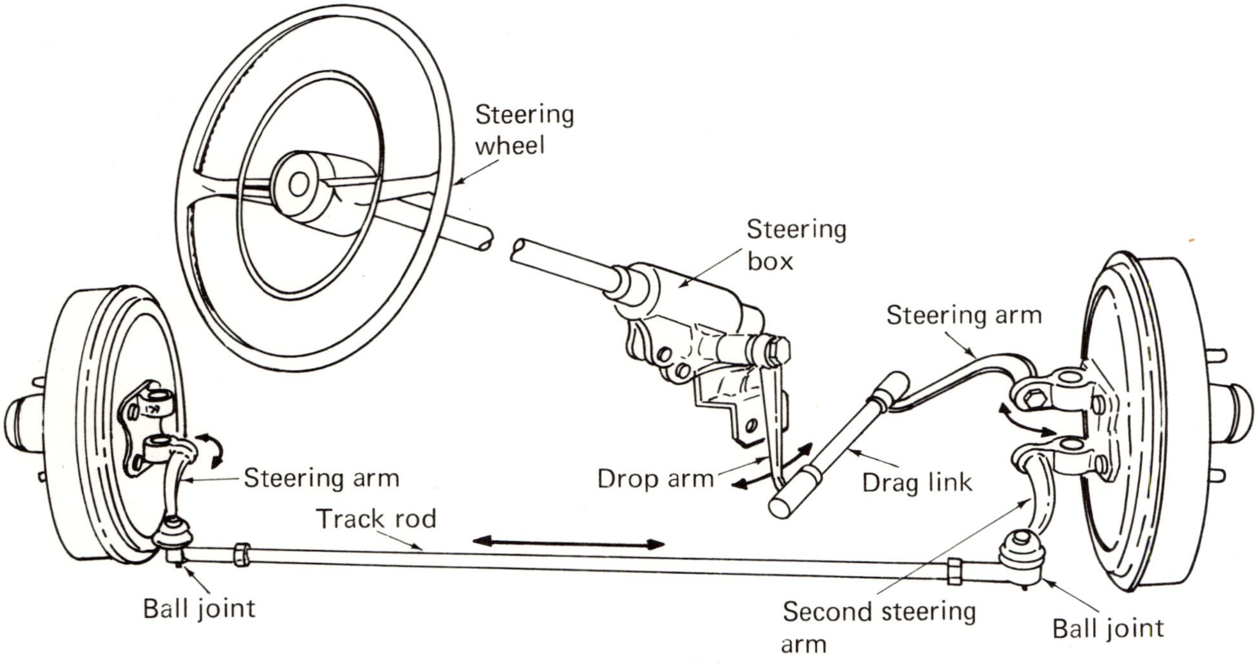

Steering wheel
Steering box
Steering arm
Steering arm
Drop arm
Drag link
Track rod
Ball joint
Second steering arm
Ball joint

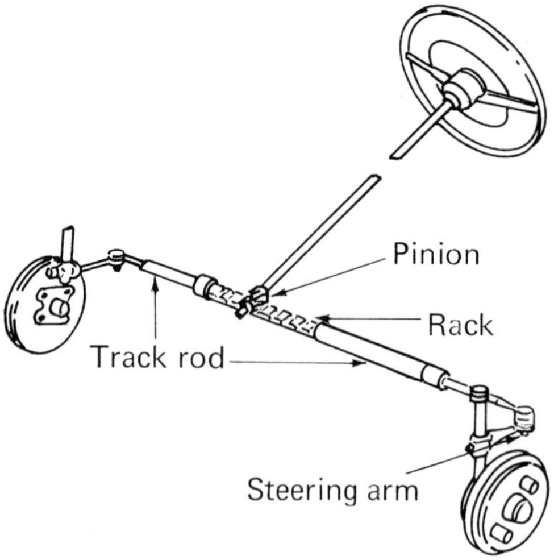


*Rack and pinion steering gear*

to the inner wheel. The steering mechanism must absorb road shock well enough so that the steering wheel does not fly out of the driver's hands when the car goes over bumps. There should also be reversability; that is, a tendency for the steering wheel to return to forward position upon release following a turn.

• *Linkages.* There are many types of steering linkages. Among these are rack and pinion, and lever arm. Steering gearbox arrangements include worm and roller, recirculating ball, and cam and lever.

• *Rack and Pinion.* This is a simple but effective mechanism and was used in many early automobiles. In the 1960s and 1970s, it achieved new popularity. The linkage between the *steering column* that leads from the steering wheel to the *track rod* or *tie rod* linkages, to which the wheels are attached, is the most direct of any steering system. The directness results in good "feel," and the action is quite responsive to the driver.

The system is a *pinion gear* at the end of the steering column that meshes with the rack gear on the tie rod. The tie rods swing at a connection with the steering arm, which may or may not have ball joints to transmit force to the wheels.

• *Lever Arm.* This occurs in a wide variety of mechanical detail. Basically, it is a system of levers that reduces the turning of the steering wheel and supplies mechanical advantage to the motion of the wheels. There is, typically, a geared steering box. The linkage may be an output shaft from the steering box of a *drag shaft*, and then to a jointed connection on the tie rod or to one of the wheels. Steering arms from the wheels articulate with *ball joints* on the ends of the tie rods.

• *Steering Boxes. Cam and Lever.* A crew-type cam is at the lower end of the steering column. The cam has a continuous spiral groove, which engages with the studs on output shafts that link with a *pitman arm* to the tie rods. This arrangement can provide constant or variable ratios.

• *Recirculating Ball.* This is also called the *screw and ball nut system.* A part of the gear on the output shaft is in constant mesh with a rack milled on one side of a nut. The nut rides on a screw at the end of the steering column. Ball bearings are between the threads on the screw, and they drive the nut along the threaded steering shaft. The balls reduce friction between the screw and nut and make steering effort easier. The "feel," however, is not as positive as with rack and pinion.

• *Worm and Roller.* A spiral cam is at the end of the steering column. As the cam moves up and down, it engages with a stud or liner that is linked through an output shaft to the rest of the steering linkages.

• *Power Steering.* Most larger American cars are fitted with power steering. Basically, some of the engine power is used to move the steering linkages. The driver's movement of the steering wheel opens and closes valves and other controls in the power steering unit. This is especially appreciated when trying to park a large car in a small space.

There is, of course, a wide variety of design in power steering units. But almost all of them operate hydraulically. The pres-

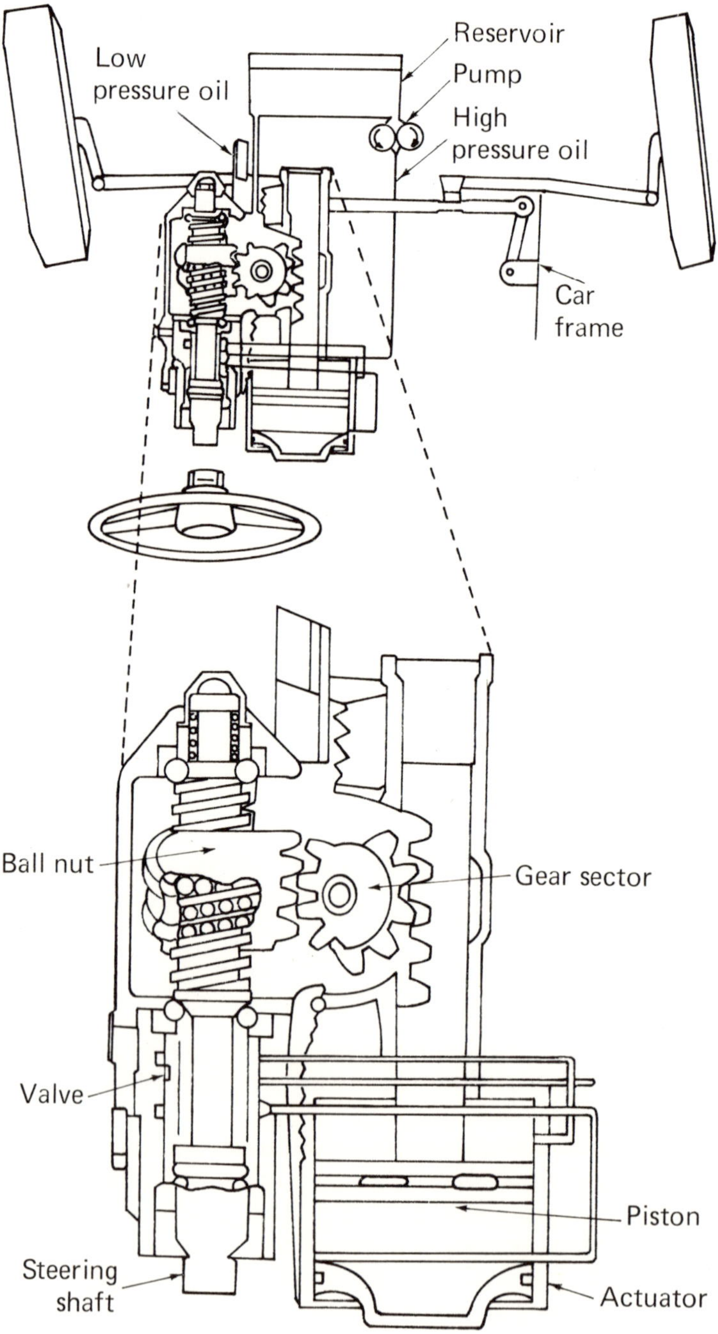

Reservoir
Low
pressure oil
Pump
High
pressure oil
Car
frame
Ball nut
Gear sector
Valve
Piston
Steering
shaft
Actuator

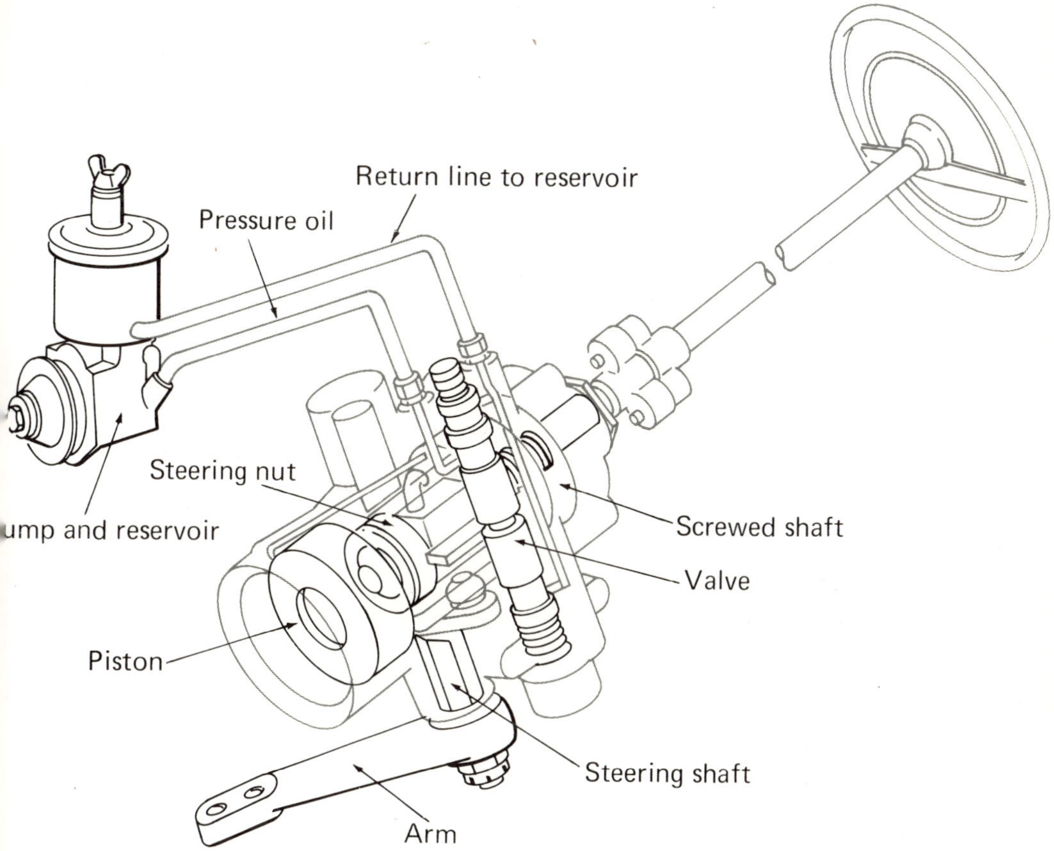


*Left and above, a power steering system*

sure to hydraulic oil is provided by a pump that is driven by a belt and pulley running off the crankshaft pulley.

The main components of a power steering system are a *power activator*, *control valve*, and a *source of power* (pump and engine). There are also steering gear boxes and linkage systems as with conventional steering. The system of components and actions in power steering is referred to as a *servo loop*. It can be placed almost anywhere in or near the steering system. Some are in the steering shaft, the linkages, or in both shaft and linkage.

In any system of any mechanical arrangement, the pump sends the hydraulic fluid under pressure to a valve. When the steering wheel is turned, the valve opens and allows oil into the activator. In the activator is a piston that activates the steering gear and, ultimately, the steering linkage. The valve also responds to bumps in the road that tend to wiggle the wheels. Oil returns to the fluid reservoir through a line from the steering gear assembly.

# suspension systems

Suspension systems are designed not only to give passengers a more comfortable, but also a safer, ride. An automobile that was nothing more than a board on wheels propelled by an engine would bounce all over the road, and, sooner or later, would probably turn over. Suspension systems have two basic functions — to absorb road shock and to improve the stability of the moving vehicle.

Springs of various types are a part of every automobile suspension. Coil springs are usually mounted in the front, and leaf springs in the rear. The axles are also part of the suspension system. They support the weight of the body and are constructed to "give" with the road bumps. Most cars have *independent front suspension*. Some also have independent rear suspension. Some are equipped with *torsion bar* and *air* suspension systems.

The basic idea in independent front suspension is that each front wheel is linked to the body of the car by a system of levers that has its own spring. Each wheel responds to bumps by itself. If the right wheel goes over a bump, the left wheel will not be

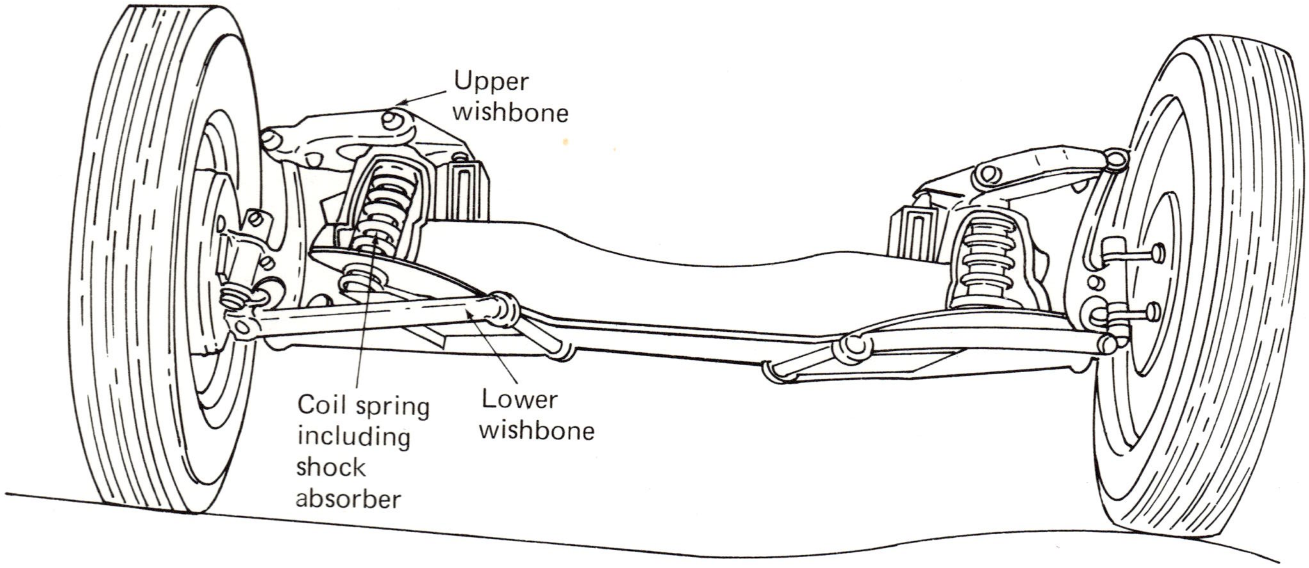
Upper
wishbone
Coil spring
including
shock
absorber
Lower
wishbone

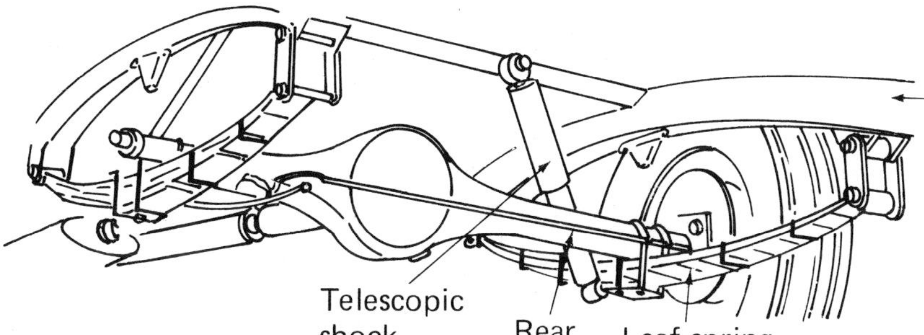


*Top shows a conventional rear suspension system;*
*left, a front suspension showing coil and wishbone system.*

affected. A common type of independent suspension is the "wishbone" system. The wheel is carried on a short axle on a T-shaped *kingpin*, which is carried on a ball joint by wishbone links extending from the chassis. A coil spring is between the wishbones. Almost all systems incorporate *shock absorbers.* The shock absorber is a hydraulic piston that absorbs energy from the springs. They also act to reduce the tendency of the car to sway on the springs. Some automobiles also have *anti-sway bars.*

Independent rear suspension is more difficult, since the axles carry torque to the wheels. Rear wheels may also be suspended on a wishbone arrangement, but the wheels must be driven by shafts equipped with universal and sliding joints. Other types include the *swing axle.* The axle is jointed and can swing up and down in response to road shocks.

In a well-designed suspension system, most of the road shock energy is absorbed before it reaches the passengers. The wheels will bounce up and down, but the body of the car should not. The design of suspension systems is a compromise between comfort and road handling characteristics. In general, a "soft" suspension will provide more comfort but less stability on curves and turns. "Hard" suspensions, usually found in sports cars, provide greater stability and more responsive turning characteristics.

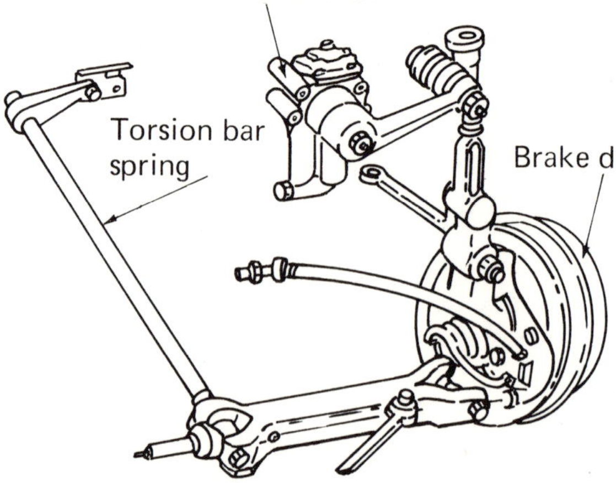


A sports car may safely be able to negotiate a turn at speeds that would cause a larger family sedan to sway and possibly "spin out."

Torsion bars act as springs that are flexed by twisting about the axis. The bars are mounted lengthwise and are usually rigidly fixed to the car frame at the rear. They are free to rotate at the front end. Stiffness is provided in the direction of the bar, which helps improve stability.

Air systems incorporate air-filled bellows that act as springs. The vehicle is supported on a cushion of air. Air pressure is maintained by a compressor that delivers air through tubes to the bellows. Various valves and sensor controls maintain the proper pressure in the bellows. Air systems are used mostly on large buses.

# future developments in automotive engineering

• *The Wankel Engine.* This is not exactly a new development, but it was not until the 1970s that there was any indication that this engine would be used on a wide scale. The greatest advantage of the Wankel engine is that rotary motion is produced in the engine. There are no pistons moving up and down.

The development of the Wankel engine began in 1926, when a German engineer, Felix Wankel, began to investigate the possibilities of rotary engines. The advantages of an engine whose motion was rotary rather than reciprocating were seen well before this time, but several engineering problems had to be solved before such engines could be built. Provisions for intake and exhaust valves or ports had to be devised. Since there were no pistons, a sealing method other than piston rings had to be devised, and this proved to be one of the most difficult problems.

The operation of the Wankel engine is simple in concept. The engine is basically a somewhat triangular rotor that revolves

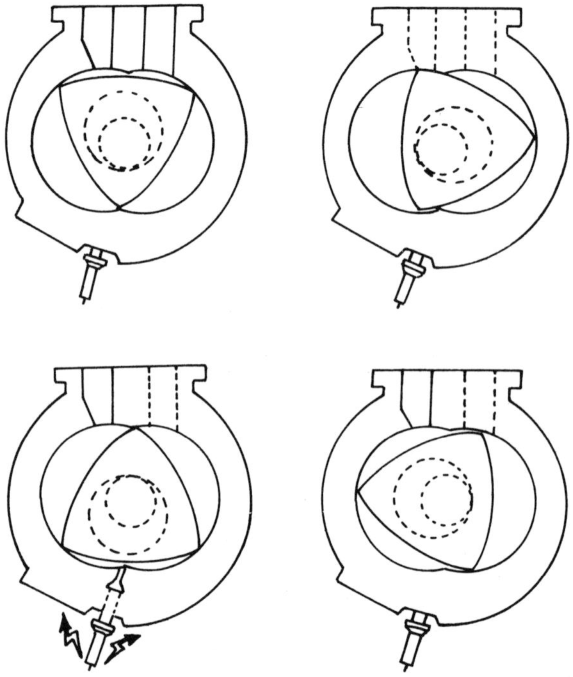

*The Wankel engine*

in a housing. The housing has a peculiar gently curved shape called epitrochoid. Seals are located at the apexes of the triangle. The seals have the effect of closing off the interior of the housing into three "chambers." There are intake, compression, ignition, and exhaust phases of the cycle as in the piston engine.

Refer to the accompanying diagrams to follow the description of the Wankel engine cycle. At the "chamber" marked 1 is the intake stroke. As the rotor moves in the direction indicated by the arrow, the volume becomes greater and the resulting partial vacuum causes fuel-air mixture to enter the chamber. The fuel

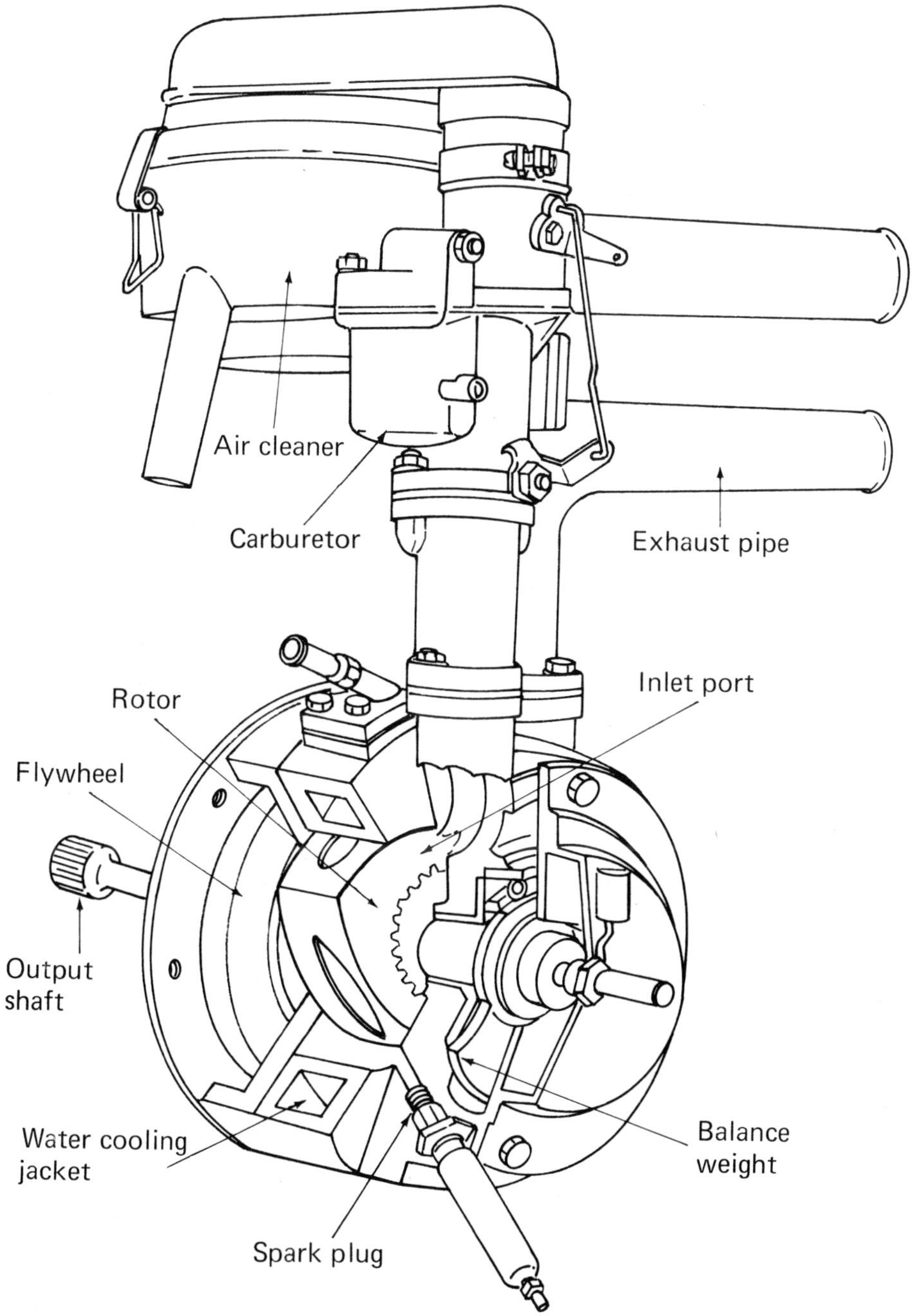

Air cleaner
Carburetor
Exhaust pipe
Inlet port
Rotor
Flywheel
Output
shaft
Water cooling
jacket
Spark plug
Balance
weight

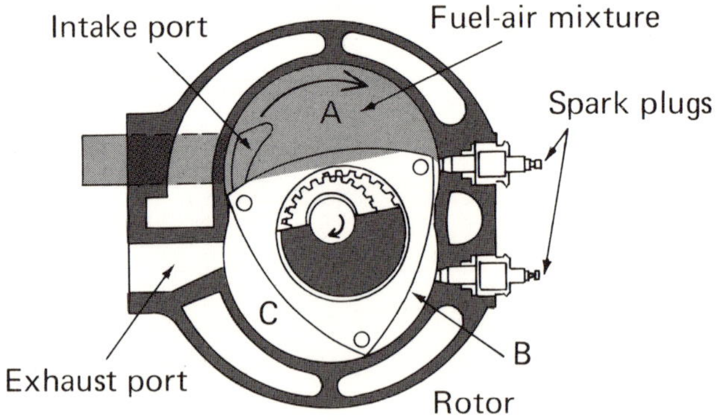


1 Intake

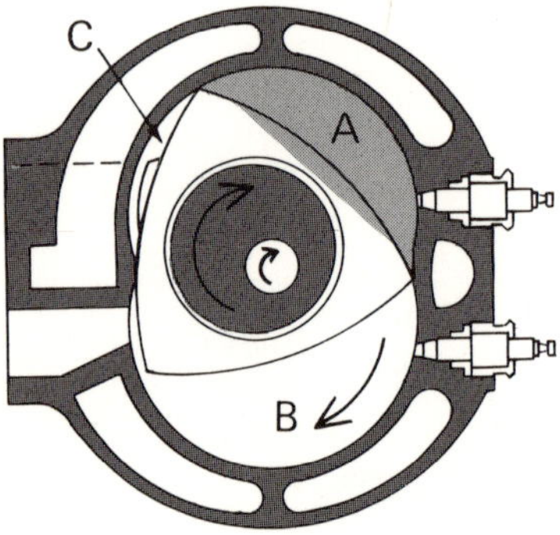


2 Compression

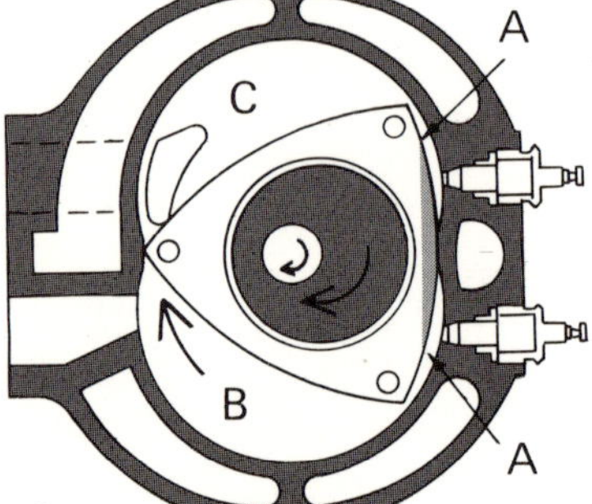


3 Ignition

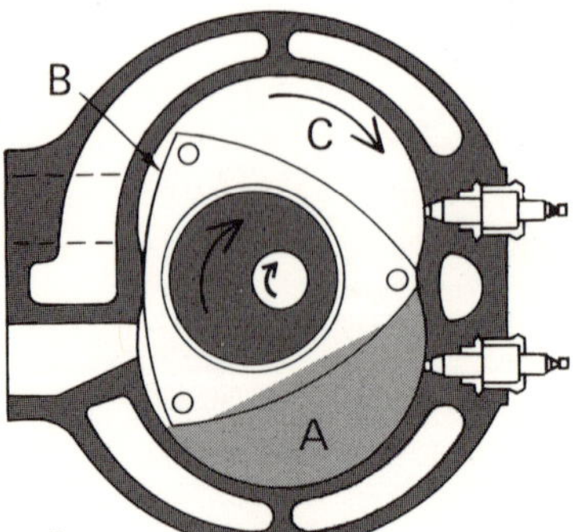


4 Expansion

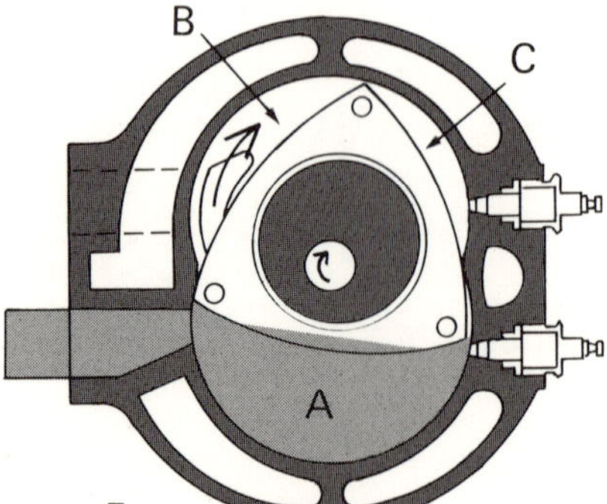


5 Exhaust

and air enter the intake port, which is always open. As the rotor continues to move in the direction indicated, the volume of the chamber decreases and the fuel is compressed. Ignition occurs at the point of greatest compression as shown. The expanding gases resulting from the combustion push the rotor over toward the exhaust port. As the rotor sweeps past the exhaust port, spent gases are expelled. All of these actions take place continuously. There is always an intake in the chamber marked 1, compression in 2, ignition in 3, and so on.

The rotor is in direct contact with an eccentric shaft. The crankshaft turns with the rotor and transmits torque to a gearbox and thence to the wheels. The rotor is maintained in the proper orientation with the housing through an internal gear that meshes with a gear fixed to the housing. These gears do not transmit any motion to the crankshaft. The crankshaft makes three rotations for every one rotation of the rotor. There is one power stroke for each rotation of the rotor. The Wankel engine, therefore, uses its displacement volume twice as often, for each cycle, as the four-stroke piston engine. A Wankel engine produces about twice as much horsepower for each pound of engine as a four-stroke piston engine. Wankel engines with more than one rotor can be mounted with the rotors in tandem.

Since the Wankel does not take up as much room as a piston engine of comparable horsepower, automobiles with this engine have more room for passengers and cargo. The engine is low and the hood can slope downward, affording better visibility for the driver. Wankel engines run with much less vibration than do piston engines. They have less than half as many parts as a piston engine, and in mass production could be less expensive than piston engines of the same horsepower. A typical Wankel engine car would have the engine, transmission, and final drive mounted transversely with front-wheel or rear-wheel drive. They can be water or air cooled and equipped with manual or automatic transmissions.

Sealing is a critical problem. In addition to the apex seals there are side seals on the rotor that fit against the internal wall of the housing. The apex seals present the greater problem. These seals are very important since they seal off the interior of the engine into the chambers. If the seals fail, there will be no compression and consequently the combustion will be ineffective in producing motion. The seals must be tight, yet at the same time

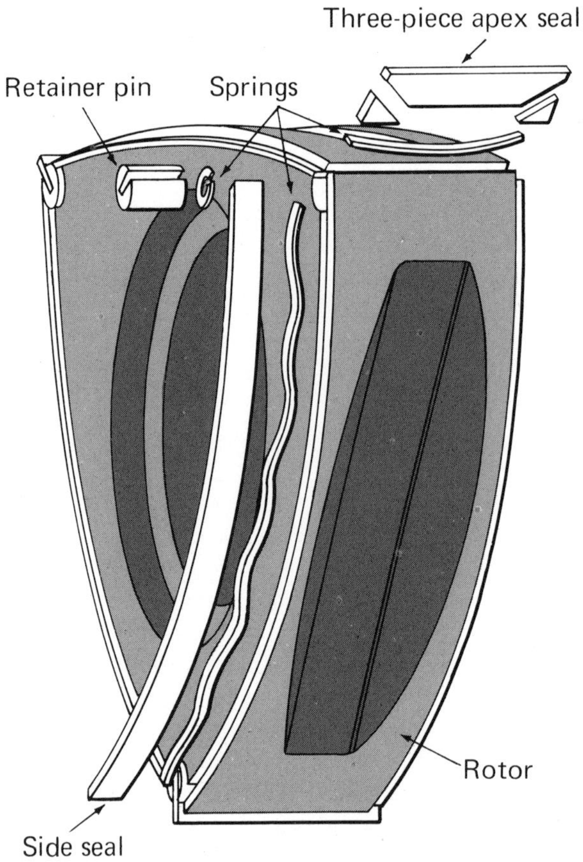


they must not interfere with the motion of the rotor. They are typically mounted on a spring set in grooves at the apex. The springs maintain the proper pressure to effect sealing without retarding the motion of the rotor. The seals must be made of a heat-resistant metal that is also resistant to wear, but not so much so that they will wear a track in the inner wall of the engine housing. The seals wear and have to be replaced. Just how often they have to be replaced is a matter not yet completely resolved. Seal replacement is a major job that involves dismantling the engine. A major aspect of research on Wankel engines will no doubt be the development of more effective and longer-lasting seals.

Wankel engines appear to run well on unleaded, low-octane fuel. But slow combustion and leakage past the seals contribute to low fuel economy. Mounting two spark plugs rather than one at the ignition "chamber" improves the economy situation. The leakage and slow combustion cause the Wankel engine to be a rather heavy emitter of pollutants. This problem can be solved with modified mufflers. Such mufflers could be equipped with afterburners or catalytic systems.

The small size of the Wankel engine and the consequent space available for emission-control devices is probably the most important factor in the adoption of this engine by an increasing number of automobile makers. Most automobile manufacturers have been reluctant to switch to rotary engines since such a change would involve expensive retooling and redesigning of production lines. However, the pressure to meet emission standards is overcoming these considerations, and in the not-too-distant future rotary engines may well outnumber piston engines in automobiles.

• *Gasoline Turbines.* Turbine engines also produce direct rotary motion. A turbine is essentially a system of many "windmills" that are turned by the expanding gases of combustion. They were tried on an experimental basis in the 1960s but were never adopted on a large scale for passenger automobiles. Cooling and muffling are particular problems with turbines. They are used on some heavier vehicles, such as trucks and buses. A number of racing cars equipped with turbines competed in several races in the late 1960s and early 1970s, but they all broke down before completing the races, although the breakdowns did not involve the turbine principle itself. Research in turbine design continues.

• *Back to Steam and Electric?* Air pollution caused by millions of internal-combustion engines is an issue of increasing concern. In recent years several prototypes of electric cars have been made. They have much more efficient batteries than the electric cars of the early part of the century. The batteries can be recharged by plugging them into regular house current. Designers claim that an overnight charge is more than enough for a normal day's driving. Many traffic engineers have proposed that small electric cars could relieve the daily morning and afternoon commuting rush hours. Commuters in the suburbs would drive their small electric cars to train stations where the vehicles would be driven onto automobile carrier trains. On arrival the commuters would drive off the trains to their places of work. They would have use of their vehicles in the city, but they would not add to pollution and would take up much less room in streets and parking lots or garages. However, the wider use of electric cars is dependent upon the development of lighter, more efficient batteries than are now generally available.

Prototypes of new steam automobiles have also been made. The burning of the propane or butane gas used to make the steam produces very little pollution. A steam automobile is somewhat more complex mechanically than an electric car, but the steam vehicle does not require recharging.

The future of the internal-combustion engine depends to some degree on the ability and willingness of the manufacturers to reduce polluting emissions effectively. If emissions are not drastically reduced, government and public pressure could possibly force the abandonment of the internal-combustion engine as a major source of transportation power.

# glossary of automotive terms

*Accelerator* — gas pedal.

*Alternator* — type of generator.

*Battery* — produces electricity by chemical action.

*Brake drum* — hollow metal cylinder fitted to each wheel.

*Brake shoe* — part of brake system inside brake drum.

*Brakes* — device to stop automobile.

*Camshaft* — turned by crankshaft, begins motion of valves.

*Carburetor* — engine part that mixes fuel with air, vaporizes mixture, and delivers mixture to cylinders.

*Choke* — device that controls amount of air entering carburetor.

*Clutch* — connects and disconnects engine to power train.

*Coil* — increases voltage produced by battery.

*Compression ratio* — comparison of volume of cylinder with piston at bottom of stroke to volume at high point of stroke.

*Connecting rods* — connect pistons to crankshaft.

*Cooling system* — removes heat from engine.

*Crankcase* — part of cylinder block that encloses crankshaft.

*Crankshaft* — metal rod in angled sections to which pistons are attached by connecting rods; when connecting rods push down on it, crankshaft turns.

*Cylinder* — basic part of engine that houses piston.

*Cylinder block* — main bulk of engine supported by frame.

*Cylinder displacement* — volume of cylinders at piston low point.

*Cylinder head* — top covering of cylinder block.

*Diesel engine* — four-stroke engine without spark plugs, used mainly in trucks, buses, and a few automobiles.

*Differential* — enables wheels to rotate at different speeds when car turns corners.

*Distributor* — distributes current to spark plugs.

*Drive shaft* — metal tube that transmits motion to final drive in rear axle.

*Electrical system* — battery, coil, points, distributor, spark plugs, generator, and voltage regulator; system that provides spark for igniting fuel and for operating all electrical parts of automobile.

*Exhaust manifolds* — metal passageways attached to cylinder block to collect exhaust gases.

*Exhaust pipe* — carries exhaust gases from manifolds.

*Exhaust system* — series of pipes to carry exhaust gases outside engine; consists of exhaust manifolds, exhaust pipe, muffler, and tail pipe.

*Exhaust valve* — allows exhaust gases to escape from cylinder.

*External-combustion engine* — engine in which fuel is burned outside cylinder.

*Final drive assembly* — mechanism that turns the driving wheels.

*Flywheel* — large wheel at end of crankshaft that helps to keep pistons moving between strokes.

*Fuel injection* — system that squirts precisely measured amounts of fuel into cylinders.

*Fuel pump* — moves fuel from tank to engine.

*Fuel system* — fuel tank, fuel line, fuel pipe, and carburetor.

*Gasket* — sealer such as between cylinder head and block.

*Generator* — device that continually recharges battery.

*Horsepower* — amount of energy required to carry out work at rate of 550 foot-pounds per second.

*Intake valve* — allows fuel into cylinder.

*Internal-combustion engine* — engine in which fuel is burned inside cylinder.

*Lubrication* — applying oil to metal engine parts to reduce friction wear.

*Manifolds* — pipes that conduct fuel-air mixture into cylinders and take exhaust gases out.

*Muffler* — metal casing that reduces engine noise.

*Oil pan* — holds lubricating oil.

*Piston* — engine part that moves in a cylinder.

*Piston rings* — sealers to keep piston force inside cylinder.

*Points* — make or break electric from battery coil.

*Power train* — system that transmits motion to wheels.

*Rack and pinion* — steering system that uses pinion gear at end of steering column, meshing directly with steering rack.

*Radiator* — contains water to cool engine.

*Rotary engine* — see Wankel engine.

*Shock absorber* — hydraulic piston that absorbs energy from springs.

*Spark plugs* — metal electrodes that produce spark from electricity.

*Starter* — electric motor that meshes with flywheel to start automobile.

*Strokes* — up-and-down motions of piston.

*Suspension system* — absorbs road shock and improves stability of vehicle.

*Tail pipe* — part of exhaust system that carries exhaust gases to outside.

*Transmission* — gears that transmit motion from clutch to drive shaft.

*Two-stroke engine* — piston engine with no valves, used in some motorboats, lawn mowers, and small vehicles.

*Valves* — engine parts that open and close with movement of pistons.

*Voltage regulator* — controls amount of current fed to battery.

*Wankel engine* — produces rotary rather than up-and-down motion in engine; has no pistons.

# some facts about the automobile and the industry

Approximately one out of every seven persons employed in the United States is in some form of business related to highway transportation.

Along with the convenience and pleasure it gives, the automobile through the years has been blamed for America's high accident rate, for rises in juvenile delinquency, and for providing criminals with an easy means of escape.

The growth of the oil industry in the United States is directly traceable to the needs of the automotive and related industries.

The automobile was not really invented; that is, no one individual can be named its inventor.

Some highlights in the development of the automobile:

1769 French army officer Nicolas Cugnot builds the first self-propelled road vehicle — a steam-driven gun tractor.

1801 Englishman Richard Trevithick builds the first steam carriage.

1804 Oliver Evans builds the first American self-propelled vehicle — a steam-driven dredge.

1827 Frenchman Onésiphore Pecqueur invents the first differential, enabling wheels to rotate at different speeds when turning corners.

1832 Englishman W. H. James patents the first three-speed transmission.

1860 Frenchman Étienne Lenoir builds an internal combustion engine that runs on gas.

1875 Siegfried Marcus of Vienna builds an internal combustion vehicle, now preserved in Vienna museum; may be oldest existing such engine.

1886 German Gottlieb Daimler invents one of the first gasoline-engine-powered automobiles.

1893 Charles E. and J. Frank Duryea build first American gasoline-powered automobile.

1895 Duryea brothers establish first American automobile company.

1902 First volume-production car appears, the curved-dash Oldsmobile.

1912 Electric self-starter is installed for the first time.

1923 Two firsts — the use of lacquer finishes, and four-wheel brakes become standard on production cars.

1933 Independent front-wheel suspension is introduced.

1940 Automatic transmission and sealed-beam headlights come into use.

1949 High-compression engine introduced for greater economy.

1960s and 1970s Drastic and numerous changes in design, transmissions, power brakes and steering, as well as safety and emission controls.

Probably the oldest existing self-propelled American car is in a private collection in East Greenwich, Rhode Island. This steam carriage built by Richard Dudgeon in 1867 ran on wooden wheels.

A steam carriage built by Francis E. and Freeman O. Stanley captured the world speed record in 1906 at Ormond Beach, Florida, by going over two miles a minute.

In the early years of the twentieth century, gas in the United States cost 10 cents a gallon.

The first American automobile race was won near Chicago in 1895 by the Duryea brothers. The winning car had two cylinders.

The first car to complete a transcontinental trip was a two-cylinder Winton in 1903. Nelson Jackson made the journey from San Francisco to New York City in 63 days.

The first annual automobile show was held in New York City in 1900. Some 300 vehicles were shown, featuring Packards, Pierces, Oldsmobiles, Cadillacs, Wintons, and Locomobiles.

General Motors was organized in 1908, the same year that Henry Ford began to manufacture the famous Model T, the first car that so many of the general public could afford to buy. Models sold from $850 to $290. From 1908 until it was discontinued in 1928, 15 million model Ts (called "Tin Lizzies") were sold.

Walter P. Chrysler was a railroad worker before he organized his automobile company; Henry Ford was a watchmaker; John and Horace Dodge sold bicycles.

Up to 1919, 90 percent of all cars manufactured were open models; after 1929, 90 percent were closed.

Well over 3,000 types of automobiles have been made in the United States since the beginning of the industry.

# for further reading

If the automobile holds a certain fascination for you, the following books may be of interest:

Bergere, Thea and Bergere, Richard. *Automobiles of Yesteryear.* New York: Dodd, Mead, 1962.
Brooks, Lester. *Great American Autos.* New York: Scholastic Book Services, 1972.
Buehr, Walter. *Automobiles, Past and Present.* New York: William Morrow, 1968.
Butterworth, W. E. *Wheels and Pistons: The Story of the Automobile.* New York: Scholastic Book Services, 1971.
Clifton, Paul. *Fastest Men on Earth.* New York: John Day, 1966.
Edmonds, I. G. *Hot-Rodding for Beginners.* Philadelphia: Macrae Smith, 1970.
Engel, Lyle K. *The Complete Book of Nascar Stock Car Racing.* New York: Scholastic Book Services, 1971.

——— *One Hundred and Thirty-Two of the Most Unusual Cars That Ever Ran at Indianapolis.* New York: Arco Publishing, 1970.

Fox, Charles. *Great Racing Cars and Drivers.* New York: Grosset and Dunlap, 1972.

Granatelli, Andy. *They Call Me Mister 500.* Chicago: Henry Regnery, 1969.

Harding, Anthony. *Car Facts and Feats.* New York: Doubleday, 1972.

Hebb, David. *Rally Book.* New York: Hawthorn Books, 1973.

——— *Wheels on the Road.* New York: Macmillan Publishing, 1966.

Hertz, Louis H. *Complete Book of Building and Collecting Model Automobiles.* New York: Crown Publishers, 1970.

Hill, Frank E. *Automobile: How It Came, Grew, and Has Changed Our Lives.* New York: Dodd, Mead, 1967.

Jackson, Robert B. *Classic Cars.* New York: Henry Z. Walck, 1973.

Jamison, Andrew. *Steam Powered Automobiles.* Bloomington, Ind.: Indiana University Press, 1970.

Lent, Henry B. *Automobile-U.S.A.: Its Impact on People's Lives and the National Economy.* New York: E. P. Dutton, 1968.

Lurani, Giovanni. *The History of the Racing Car.* New York: T. Y. Crowell, 1972.

Musciano, Walter A. *Building and Operating Model Cars.* New York: Funk and Wagnalls, 1970.

Posthumus, Cyril. *Land Speed Record.* New York: Crown Publishers, 1972.

Stambler, Irwin. *Unusual Automobiles of Today and Tomorrow.* New York: G. P. Putnam's, 1972.

Stevenson, Peter. *The Greatest Days of Racing.* New York: Charles Scribner's, 1972.

# index

Accelerator, 29
Afterburners, 52
Air pollution, 6, and emission control devices, 50, 52, 99
Alternator, stator of, 41
Armature. *See* generator.
Auto mechanics, 1-2
Automobile, history of, 3-5
Axle(s), 89, 91

Battery (-ies), cells of, 33; in electric cars, 5; recharging of, 38-39, 41; voltages (def.), 33, 35
Benz, Karl, 4
Block. *See* cylinder block.
Brake(s), 74-75, 77-81; brake fluid, 75, 77-78, 81; brake pedal, 75, 78; brake shoes, 75, 78; definition of, 74; disc type, 77-78; drum type, 75, 77; dual, 75; fading, 74; hydraulic system, 74; power type, 78, 81
Breaker points. *See* distributor.

Cam and lever (steering gearbox), 85
Camshaft, 20, 55; description, 17; and oil pump, 44
Carburetor, 24, 27; automatic choke, 30; barrels of, 29-31; choke valve, 30; float cham-

ber, 28-29; flooding, 32; purpose of, 28; throat, 28; throttle, 29-30; *venturi,* 28-29
Catalytic convertor(s), 52
Clutch, 56, 59; types of automatic, 65; fluid, 65, 66; plates of, 59
Coil. *See* ignition coil.
Combustion chamber, 11, 13-14, 18, 24, 27, 30
Commutator. *See* generator.
Compression (stroke), 9
Condenser (capacitor). *See* contact points.
Connecting rod(s), 11, 13, 55
Contact points, 33, 35-37; condenser in circuit of, 35
Cooling systems, air, 46-47; and antifreeze, 49; and pressure cap, 48-49; thermostat, 47; water, 47-49; and water pump, 47
Crankcase, 16
Crankpins, 13
Crankshaft, 11, 13, 16-17, 21, 23; crank handle for, 41; in splash system lubrication, 43; in water cooling, 47; in Wankel engine, 97
Cuguot, Joseph, 3
Cut-out device. *See* generator.
Cylinder(s), 9, 11, 14, 16, 18, 22; arrangement of, 13; cylinder displacement, 23; firing sequence, 13; pistons in, 55; valves of, 7; water jacket around, 47
Cylinder block, 14, 16
Cylinder head, 14, 18

Daimler Benz Co., 4
Daimler, Gottleib, 4-5
De Rivaz, 3
Diesel engine(s), 31; uses of, 55
Diesel, Rudolf, 55
Differential, 71; non-slip type, 72
Distributor, 33, 35-37; breaker points, 36; definition, 36; spark advance mechanism, 37
Distributor cap, 36
Downdraft carburetor, 28
Drive shaft, 67-68
Drive train. *See* power train.
Driven member. *See* fluid flywheel clutch.
Driving member. *See* fluid flywheel clutch.
Dynamometer, 23

Electric cars, 5
Electrical system, 33, 35-39, 41-42
Emission control devices, 52, 99; and lead in gasoline, 23
Energy of motion, 4
Engine(s), compression ratios of, 22-23; external combustion, 4; four-stroke cycle, 9, 11, 13, 53-54; horsepower capacity of, 23; internal combustion, 3-5, 41; overhead valve, 18, 20; principles of gasoline, 7, 9, 11, 13; short stroke, 13; structure of, 14, 16. *See also* diesel, gasoline, steam, two-cycle, and Wankel engines.
Engine housing, for Wankel engine, 93-94
Exhaust manifold, 16, 50
Exhaust pipe, 50

Exhaust stroke, 9
Exhaust systems, 11, 50-52, 55; and air pollution, 52; dual-, 50; waste gases, 4

Fan belt, 47
Final drive, 67-68
Flick-over mechanism. *See* fuel pump.
Float chamber. *See* carburetor.
Fluid flywheel clutch, 65-66
Flywheel, 55; attachment to clutch, 59; attachment to crankshaft, 11, 13; ring gear around, 41-42
Flywheel plate. *See* clutch.
Four-stroke engine(s), 9, 11, 13, 53-54
Four wheel drive, 73
Front wheel drive, 72
Fuel(s), electric power, 5; gasoline, 2, 4, 23; hydrogen and air, 3; with higher compression ratios, 22; octane ratings of, 22-23; steam, 3-5; for Wankel engines, 99
Fuel injection system, 27, 31-32; definition, 32
Fuel line, 24, 26-28
Fuel pump, 24, 26-28, 32; diaphragm, 27-28; flick-over mechanism, 28; in fuel injection, 32; mechanical/electric, 27-28; outlet valve, 28
Fuel system, definition, 27; distribution to combustion chamber, 24, 26, 32; gravity feed, 24; in Wankel engine, 94-95
Fuel tank(s), 24, 26-27

Gasket(s), 14, 16
Gasoline engines, 4-5, 32
Gear(s), in automatic transmissions, 65; definition of, 59; gearbox, 60-61, 64, 67, 97; planetary, 67; ranges of, 61, 63; ratios, 60; speeds, 60-61; synchromesh, 63-64
Gear ratios, 60
Gear shift lever, 61
Gearbox. *See* transmission and gears.
Generator, 33, 38-41; armature of, 38-39; brushes, 39; commutator, 39; cut-out device, 39, 41; principles of, 38-39; voltage regulator, 39, 41

Horsepower, 23
Hydraulic brake systems, 74, 87; master cylinders of, 75
Hypoid. *See* ring and pinion gear.

Ignition coil, 33, 35-36; windings of, 35
Ignition system, 33, 42, 55; in Wankel engine, 97
Impeller. *See* torque converter.
Intake manifold, 16, 29, 32; vacuum in, 78
Intake stroke, 9

Lever arm steering, 85
Lubrication, 14, 16; of cylinder in engine, 11, and oil pump, 44-45; sump, 43, 45; types of, 43

Main bearings, 13

Marcus, Siegfried, 4
Master cylinder, 75
Mechanical motion, rotary, 11, 93, 99; reciprocating, 11, 93
Mercedes-Benz, 4
Muffler(s), 50, 52, 99; baffles on, 52; straight-through, 52
Multiplication gearing, 60

Octane ratings. *See* fuel.
Oil (types of), 45
Oil filter(s), 45
Oil pan, 14, 16
Oil pump; gear type, 44; pressure relief valve, 45
Otto cycle, 9
Otto, Nicolaus, 4, 9
Outlet valve. *See* fuel pump.

Paris-Bordeaux race, 5
Panhard-Lavassor, 5
PCV (positive crankcase ventilation), 52
Pinion gear, and final drive, 71; and rack and pinion steering, 84; and starter motor, 41-42
Piston(s), 7, 9, 11, 17, 22-23
Piston pin, 13
Piston ring(s), 11
Ports, intake and exhaust, 53, 93-94, 97
Power steering, 85, 87-88; servo loop, 87
Power train, 56, 59-61, 63-68, 71-73
Pressure plates. *See* clutch.
Push rod, 20-21

Rack and pinion steering, 84
Radiator, 47-48
Rear axle assembly, differential, 71; final drive, 67-68; ring and pinion gear, 67-68
Recirculating ball (steering gearbox), 85
Reduction gearing, 60
Revolutions per minute (rpm), 23
Ring and pinion gear, 71
Ring gear. *See* flywheel.
Rocker-arm assembly, 20
Rotor. *See* distributor.

Shock absorber(s), 91
Single dry plate friction. *See* clutch.
Society of Automobile Engineers (SAE), 45
Solenoid, 42
Spark advance mechanism. *See* distributor.
Spark plugs, 33, 36, 55, 99; gap, 35-36; in cylinder head, 14
Spiral bevel. *See* ring and pinion gear.
Spring(s), 89, 91
Starter, 3, 5; definition, 41
Starter switch circuit, 42
Steam engines, 3-5
Steering, 82, 84-85, 87-88; early development, 82; gearboxes, 84-85; linkages, 84-85
Step-up transformer. *See* ignition coil.
Sump, 14, 16, 44-45
Suspension systems, 89, 91-92; air, 89, 92; independent front suspension, 89, 91; shock absorbers, 91; torsion bars, 89, 92

Synchromesh. *See* gears.

Tail pipe(s), 50
Thermosiphon (cooling system), 47
Thermostat, 47
Throttle (valve). *See* carburetor.
Throw-out bearing. *See* clutch.
Torque, 97; definition, 59; impeller, 66; stator blades, 66; torque converter, 65-67; turbine, 66
Transmission, automatic, 64-67, 97; clutch types used in, 65; gearbox in, 67; gear ratios, 67; torque converter, 65-67
Transmission, manual, 59-61, 63-64, 97; gear ranges of, 61, 63-64; gearshift lever, 61, 64; gear speeds, 60-61; overdrive, 61
"Tune-up," 36
Turbine(s), 4, 66

Two-stroke engines, 53-55

Universal joint, 67-68; constant velocity, 72

Valve(s), 14; arrangement of, 17-18; poppet type, 17; valve train for, 20-21; intake and exhaust, 7
Valve lifter, 20; hydraulic, 21
Valve train, 20-21
*Venturi*. *See* carburetor.
Voltage. *See* battery.
Voltage regulator, 33, 39, 41

Wankel engine, 93-94, 97-99; advantages of, 97, 99; engine cycles, 94, 97; sealing, 94, 97-99
Wheel(s), attachment to steering mechanism, 82, 84-85; brake attachment to, 75
Worm and roller (steering gearbox), 85